THE ENERGY GLUT

About the authors

IAN ROBERTS is professor of public health at the London School of Hygiene and Tropical Medicine, Britain's national school of public health and a leading postgraduate institution for research and education in global health. His main research interests are in the prevention and treatment of serious injuries and in the links between energy use, sustainability and health. He trained first as a paediatrician working in intensive care in the UK and then in public health at the University of Auckland, New Zealand and McGill University in Canada. He now works on large-scale international clinical trials to find better treatments for seriously injured patients.

PHIL EDWARDS is a senior lecturer in statistics at the London School of Hygiene and Tropical Medicine and head of the Department of Nutrition and Public Health Intervention Research. He trained in mathematics and statistics at the University of Warwick and teaches statistics across the Master's programme. His main research interests are in transport and health, in particular road-traffic injury.

THE ENERGY GLUT

Climate Change and the Politics of Fatness

***

IAN ROBERTS
WITH PHIL EDWARDS

Zed Books
LONDON & NEW YORK

The Energy Glut: Climate Change and the Politics of Fatness was first published in 2010 by Zed Books Ltd, 7 Cynthia Street, London N1 9JF, UK and Room 400, 175 Fifth Avenue, New York, NY 10010, USA

www.zedbooks.co.uk

Designed and typeset in Monotype Bulmer
by illuminati, Grosmont
Index by John Barker
Cover design by www.alice-marwick.co.uk
Printed and bound in Great Britain by
CPI Antony Rowe, Chippenham and Eastbourne

Distributed in the USA exclusively by Palgrave Macmillan, a division of St Martin's Press, LLC, 175 Fifth Avenue, New York, NY 10010, USA

A catalogue record for this book is available from the British Library
Library of Congress Cataloging in Publication Data available

ISBN 978 1 84813 517 8 hb
ISBN 978 1 84813 518 5 pb
ISBN 978 1 84813 519 2 eb

Contents

The fat planet

I am in Bogotá airport, in a fast-food restaurant that could be a fast-food restaurant anywhere in the world. On the table in front of me is a sheet of thin green paper, a place mat that doubles as an advert for a range of calorie-laden delights, including Jakobos and Monchers, which would appear to be deep-fried chicken nuggets impregnated with a bolus of bright yellow cheese. What I thought might be the least energy-dense option came with a jagged mound of fries. These are now calling to me from my otherwise cleaned plate, which I have shoved to the other side of the table so as to be as far away from me as possible. Although I still have nine hours of seating and eating ahead of me on my flight, I am already sure that my personal energy balance is in unhealthy credit.

Looking out over Bogotá airport, an enormous playground of dun-grey concrete squares etched with yellow and white lines in curves and tangents, dotted with stubby trucks with tractor wheels dragging huge shark-nosed planes behind them, it strikes me that airports must be the most energy-dense environments on earth. Bright red tankers pump fossil fuels through thigh-thick hoses into the wings of waiting aircraft, while shiny steel catering containers are loaded like military coffins into their bellies. Their calorific contents will be mechanically exhumed shortly after take-off to feed the rows and rows of neatly polarized, functionally paralysed, seated passengers.

I am on my way back to London after a month in Colombia visiting the trauma hospitals taking part in an international clinical trial (the CRASH 2 trial) that we hope will provide doctors with an effective treatment for life-threatening bleeding, most of which is caused by car crashes and violence. Most of the thinking for this book was done in Colombia, early in the morning, on bus journeys between cities, or late at night, while winding my way home from the emergency department of Hospital Universitario San Vicente de Paúl in Medellín. I have spent most of my medical career treating or researching energy-related health problems, but in recent years I have come to see they are nothing more than different manifestations of the same basic planetary malaise.

Worldwide, over a billion adults are overweight and 300 million are obese. One-third of the US population is obese. Government scientists predict that by 2050 more than half of the UK population will be obese, making the UK 'a predominantly obese society'. Australians, Argentines, Belgians, Bolivians, Canadians and Chinese – everyone is getting fatter. This will affect our health and our well-being, increasing the risk of diabetes, heart disease, stroke and cancer and taking some of the joy out of living. At the same time, global warming is the greatest environmental threat the planet has ever faced. Climate scientists predict an average global temperature increase of between 1.5°c and 6°c by 2100, depending on the extent of future emissions, possibly reaching a temperature not experienced in the past 100,000 years. This will have dire consequences for plant and animal life and for our health. Even cautious scientists are talking in apocalyptic terms about famine, disease and environmental refugees. That climate change is real and man-made is not in doubt. All that is uncertain is how bad it will turn out to be.

If you think that obesity and climate change are unrelated, you are wrong. The human race is getting fatter and the planet is getting hotter, and fossil fuels are the cause of both. This book is the story of energy. The main characters are sunlight, petroleum, movement, food and fat, with money and greed as minor characters albeit of major importance. The story begins with sunlight and decay and will almost certainly end with sunlight and decay unless we act now.

About 150 million years ago, microscopic plankton suspended in the ancient seas sipped carbon dioxide from the surface waters and with the energy from sunlight produced simple sugars and oxygen in a chemical reaction called photosynthesis. In doing so they had converted light energy into chemical energy. The carbon they had absorbed was incorporated into their biological structures, in the same way that the food we eat becomes part of our bodies. When these tiny organisms died, they sank to the bottom of those prehistoric oceans, where they were covered by mud and sand. Here they rotted and, with heat from below and pressure from above, their carbon-containing remains were cooked into a thick black sludge called petroleum. In energy terms petroleum is sunlight jam. For millions of years this jam was safely locked away in its underground larder, but due to the slip and slide of the earth's crust some of it seeped out onto the earth's surface, oozing out of the rocks and up into salt wells.

The energy glut

Apart from the evolution of man and the ascendancy of his rapacious economic system, not much more happened until about 1850 when a savvy New York lawyer called George Bissell, whilst visiting his sick mother in New Hampshire, dropped in on his old university professor at Dartmouth College. During his visit, Bissell spotted a small sample of 'rock oil', also known as petroleum, which had been left there by another of his professor's former students, who was working in Pennsylvania. Bissell knew that petroleum was flammable and wondered if he could make money marketing it as a fuel for lamps. He could, and he did, and for many profitable years petroleum lit up the world.

Then, in 1880, Thomas Edison invented the light bulb, and the electricity generation industry grew to power it. Electricity threatened to turn out the lights on Bissell's petroleum business, and it might have done were it not for the fact that petroleum-powered motor vehicles had begun their ascendancy. Thanks to Henry Ford and the mass production of automobiles, walking and cycling, the usual modes of human transportation, went into a tailspin, slowly but inexorably reducing the amount of bodily energy people expended getting about. This motorization of movement was more than enough to send weight and waistlines on their upward trajectory, but a few decades later petroleum found a niche in our food system and our fat fate was sealed. In the 1940s, petroleum ignited an agricultural revolution that resulted in massively increased food yields. The energy intake of the population became higher than its energy output, and, although the imbalance was slight, the population of planet earth started getting fatter.

The increase in the combustion of petroleum released in huge quantities the carbon dioxide that those ancient microorganisms had absorbed from the Jurassic oceans millions of years previously. And as it built up in the atmosphere, the blanket of carbon dioxide trapped the heat of the sun, steadily warming the earth.

The law of the conservation of energy states that energy can neither be created nor destroyed, only changed from one form to another. In this story we see light energy from the sun captured as carbohydrate by plankton, then cooked and concentrated by the heat of the earth into petroleum, which is concentrated chemical energy. When burned in the engine of a car, this chemical energy is transformed into movement (kinetic energy). Of course, our food energy also comes from the sun and we are biologically programmed to fill up our tanks on a regular basis in order to keep the human body moving. But motorization has meant that petroleum has replaced food as the main energy source for human movement. The unused food energy is stored as body fat and the burning of the petroleum is warming the earth. Global fatness and global warming are different aspects of the same energy problem. A pulse of fossil-fuel energy released from the ground only as recently as the last century has propelled the average human weight distribution upwards, like an earthquake under the sea sends a tsunami towards the shore. At the same time, the carbon dioxide pollution that the burning of petroleum entails is warming the atmosphere, endangering our ecosystems. Our increase in body fat is a threat to our health. Global warming is a threat to our survival.

Neither losing weight nor saving humanity will be easy. To achieve both, each of us will need to become an energy activist and remain so for the rest of our lives. Obesity and climate change are political issues and we need to take political action in response to them. I should say now that political in this context means more than the trivial daily drama played out in the media between republicans and democrats, right wings and left wings. Fat politics is about how and why certain groups of people have made decisions that determine how much we move our bodies, what and how much we eat, and how we spend our time and money. The science and politics of obesity and climate change are accessible to everyone. The reason they are so often misunderstood is that the people who make the decisions prefer it that way. But if we succeed, we will have reasserted our dignity as human beings. Treading more lightly on our planet will make us healthier and happier and richer in every sense of the word.

Chapter 1 introduces the global 'obesity' epidemic. What do doctors mean when they say that someone is too fat and why are more and more people succumbing to this insidious but unstoppable affliction? Who should be held responsible for increasing levels of obesity? Is fatness something that we bring upon ourselves, a problem of our own making, or is it a wider social issue? Chapters 2 and 3 consider motorization and food, the main villains in the obesity drama that conspire to make us fat. We will see that far from being a personal failing, obesity is a normal human response to a sick environment, the bodily consequence of living in a world flooded with cheap energy. As a result of petroleum-powered transportation and the road danger it creates, we walk and cycle less than ever before in the history of the world and our personal energy output has plummeted. However, at the same time we are surrounded by masses of cheap food energy. Chapter 4 examines the wider economic and political stage on which the tragedy of obesity is played out. Why is fossil-fuel energy so cheap? Who are the winners and losers in an energy-dense world?

The central chapters outline the steps needed to regain control of our bodies and our destiny. Obesity and climate change are environmental problems and to tackle them we must transform the environments that we have created for ourselves. We will see that under a system of carbon rationing, achieving and maintaining a healthy body weight becomes easy. We then take a close look at the bicycle as a means of mass transportation, a revolutionary invention that will enable us to avoid obesity whilst ensuring sustainability. Next, we consider how to reclaim our streets and neighbourhoods from the lethal motor vehicle traffic that currently blights them, so that we can begin to safely move our bodies again, in the way that they were designed to be moved. Finally, we examine how we can make our home a safer place for ourselves and our families. The desire to eat is normal and part of our evolutionary heritage. We must recognize this and modify our homes accordingly.

The closing chapter offers a vision for a better future. Although it is essential that we understand why our relationship with our planet is flawed, the motivation to forge a better bond, one in which we tread lightly, is fuelled by imagination and hope, and fortunately for humanity neither is in short supply. The good news is that tackling climate change could be the next great advance in human development.

In many respects this book offers a low-energy diet for a fat and fevered planet. Most authors of diet books have years of experience helping overweight people to lose weight, and have recognized qualifications in nutritional science. Others are charlatans who exploit for financial gain the private trouble of the bulging waistline. I trained as a medical doctor, working mostly in paediatric intensive care, before moving into academic public health, which means that I teach students and do research on the causes and prevention of disease.

Public health links the private troubles in people's lives with the public issues that give rise to them. It links the personal with the political. Public health research is like reassembling a Russian nested doll, in which the smallest, innermost doll is the sick or injured patient, and the outer layers of dolls are the causes. The immediate cause of disease or injury, the doll surrounding the innermost, is the most easy to identify because it is the closest to the patient. Doctors often become strong advocates for prevention at this level because the immediate causes of disease are often obvious to them from their daily contact with patients. The senior physicians on the intensive-care unit where I worked as a junior doctor campaigned vigorously for child restraints in cars to reduce the number of crash-damaged children that we saw on our daily ward rounds. Public health doctors are also interested in these immediate causes but would also want to know why car use is increasing and why more and more children are being driven to school. This would be the next layer of causation. If there was less road traffic in general, the roads would be safer and there would be fewer injuries, whether or not seat belts were used. And it does not stop there. There may be another layer of causes beyond this.

When it comes to prevention, tackling the outermost causes of disease is often the most effective because this usually solves many health problems at the same time. Policies that encourage children to attend their local schools would decrease the amount of school-related travel and so reduce the risk of children being involved in road traffic crashes on the journey to school. They would also reduce traffic-related air pollution, which causes asthma and other respiratory problems, and would cut green-house gas emissions. The problem is that tackling the outer layer of causes is more difficult because it usually involves taking on powerful commercial and political interests.

8

The fat planet

I wrote this book because when assembling the Russian doll of road trauma I came to see that the outermost cause of road traffic injury is also the cause of obesity and the cause of the climate change that threatens our survival. As a doctor, I have seen what fossil-fuel energy can do to the human body. I once anaesthetized a ten-year-old girl, the victim of a high-speed road crash, so that she could be taken for urgent surgery to stop her internal bleeding. When she arrived at the hospital she was awake but deathly pale. I reassured her that she would be fine. She never woke up. Experiences like these scratch grooves in the memory, which later become conduits for emotional energy, directing strong emotion to issues that most people treat with indifference. I have seen what energy can do to flesh and now I see energy everywhere.

Climate change could have devastating impacts on all of humanity but many believe that cutting greenhouse-gas emissions is primarily a problem for government. On the other hand, while hundreds of millions of people all around the world are getting fatter, governments tell us that this is a lifestyle problem, and that we should take more responsibility for our weight. This book argues that climate change and fatness are different facets of the same basic problem and that everyone, individuals and governments, has a role to play in solving them.

1

Fat people and fat populations

'By the way, Peter, who will be doing your interview tonight, is himself obese. He has been struggling with his weight for years.' I am about to be interviewed on a live BBC radio programme, having published a letter in the *Lancet* medical journal about how bulging Western waistlines are forcing up world food prices and worsening climate change (Edwards and Roberts, 2008). Perhaps because the *Lancet* letter linked three topical issues, the media picked up the story and spread it around the world like a computer virus. The letter was about weight gain across the whole of society but the news media had clearly decided that obese people were the problem. 'Obese blamed for world's ills' reported the BBC. 'Do you mind me asking if you're fat, Dr Roberts?' asks Peter, the obese interviewer. 'Yes I am', I confidently reply, 'in fact we are all fat: I'm fat, you're fat, the prime minister is fat and the Queen is fat. The whole of our society is fat. Even the thin people are fat.' I was serious and this chapter explains why.

Fat people and fat populations

Peter, it would appear, had lost in his struggle with weight. But why was it such a struggle in the first place? Was his problem just a lack of discipline at the dinner table, a penchant for snacking between meals, or a strange aversion to the gym? If he laid out his weekly dietary ration on nutrition guru Gillian McKeith's table of shame, would she have been horrified by the conspicuous absence of flax and oxygen-forming chlorophyll? For decades, fat people have been treated like weak-willed social deviants as though obesity was a personal failing, the inevitable consequence of gluttony and sloth. Some people are fat, others are not. Bad luck the fat ones. You are what you eat, they say. Fat people eat too much, others do not. Fat people don't take enough exercise, others do. This is the popular view of obesity at the beginning of the twenty-first century and it is completely and utterly wrong. This chapter takes a closer look at obesity and examines just who is getting fat and why. It asks whether getting fat is a lifestyle issue – the bodily manifestation of a plethora of consumer choices freely made, as most governments would have us believe – or whether something more insidious is going on. But let's start with the body. What does being fat mean?

The human body is a vehicle perfectly designed for your personal transportation. It will keep running until the day you die. Whether moving or not it requires energy, but provided that you fill it up regularly with enough food it will meet most reasonable transportation demands. It has an important special feature. If the amount of food energy taken in is surplus to requirements, it will store the excess energy as fat. This energy can be called on later if needed, allowing the body to keep running for weeks, even on an empty tank. Body fat accumulates when the amount of energy we eat as food exceeds the amount of energy we use moving around and keeping warm. Fat is stored energy.

11

Most of our body fat is stockpiled under our skin and around our internal organs as adipose tissue, which is mostly made up of fat cells, also known as adipocytes. Fat cells contain a large droplet of lipid. This lipid, which has the consistency of margarine on a hot day, pushes the nucleus and everything else to the edges of the cell. The lipid droplet plumps up the cell, rather like a silicone implant plumps up a breast, and the bigger the droplet the fuller the cell becomes. As we get fatter, the number of fat cells in our body increases, and the fat cells that we already have expand as the lipid droplets they contain get bigger. Although used as a high-density energy store, there is nothing inert about fat. We could make an analogy between our body fat stores and the groceries on the shelves of a busy supermarket. Even as shoppers take food off the shelves, they remain well stocked, because the food taken off is continually replaced. If we stock up on energy faster than we use it up, we get fatter, and if we use it up faster than we stock up we get thinner. Energy intake need only be slightly higher than energy output for fat to accumulate. Doctors gauge how fat we are by working out our body mass index (BMI), which is our weight in kilograms divided by our height squared (height in metres multiplied by height in metres). BMI as a measure of fatness was invented 160 years ago by Adolphe Quetelet, a Belgian mathematician, as a tool to estimate a healthy body weight taking into account how tall a person is. Doctors diagnose 'overweight' if a person's BMI is between 25 and 30, and 'obesity' if it is 30 or more.

More fat people

In recent years, obesity has hit the headlines because the percentage of the population that is obese has skyrocketed. In 1994, 14 per cent of English men were obese. Ten years later the figure was 24 per cent (Foresight, 2007). Nearly one in four men in England is now obese. On the other side of the Atlantic the situation is even worse. In the USA, one-third of the adult population is obese. Obesity is now so common in the Western world that it has ceased to be newsworthy. With nearly a quarter of the population obese, the media has now redirected its spotlight onto the 'severely obese', those with a BMI of 40 or more, and the 'super obese', those with a BMI over 50. Ten years ago, severely obese people were rare creatures and it would have taken a diligent journalist to find them, but in 2004, 1 per cent of the population of England had a BMI over 40. Thanks to a growing number of sensational, although information-lean, health documentaries, with names like *Half Ton Mom*, *The World's Biggest Boy* and *Fat Doctor*, most evenings we can settle in front of the television and watch the private sufferings of the super obese paraded for our entertainment. We can look on aghast as these sad super-sized specimens are extracted from their homes by fire crews with lifting gear, loaded into ambulances and road-freighted to hospital for urgent obesity surgery.

Comparing ourselves with others who are a lot fatter probably makes us feel better. It might also make us feel thinner. Researchers from University College London compared two general population surveys, conducted eight years apart, which collected data on self-perceived weight (Johnson et al., 2008). They wanted to find out whether the media's obsession with the super obese might be giving the wrong impression about how fat one had to be to meet the medical criteria for overweight. They found that whereas the percentage of the population who were overweight (BMI over 25) had risen, the percentage of the population who considered themselves overweight, had actually fallen. In 1999,

81 per cent of overweight people correctly identified themselves as overweight, compared with 75 per cent in 2007. Other studies have shown that the ability of parents to identify correctly when their children are overweight has become worse with time (He and Evans, 2007). Has the media coverage of super obesity backfired? It would appear that as we get fatter we are less likely to think of ourselves as fat because we compare ourselves with the people around us, who are getting fatter too.

Fewer thin people

At the same time, thin people, those with low BMIs, have also been in the news. But this time the problem is that there are so few of them in the general population that those with a very low BMI stand out as being 'abnormal'. In 2000, the British Medical Association (BMA) provoked a flurry of news coverage with a report slamming the media's obsession with 'abnormally thin' fashion models (BBC, 2000). According to the BMA the degree of thinness exhibited by the models was both 'unachievable and biologically inappropriate'. The problem of skinny models resurfaced in 2006 when the Spanish Association of Fashion Designers banned models with a BMI less than 18 from Madrid Fashion Week. The British secretary of state for culture, media and sport Tessa Jowell also wanted to see thin models chased off the catwalk. She applauded the Madrid decision and urged London Fashion Week to do the same.

There is certainly evidence that being thin is becoming harder to achieve. In 1994, 7.4 per cent of English women had a BMI less than 20. Ten years later, the figure was 5.6 per cent. Over the same period, the percentage of women with a BMI under 18.5 fell from 2.2 per cent to 1.7 per cent (Foresight, 2007). It would seem that even the thinnest people in our society are slowly getting fatter. But the evidence that being thin is biologically inappropriate is less convincing. The US Nurses Health Study measured the BMIs and other health-related risk factors of 115,000 nurses and then followed them up for sixteen years (Manson et al., 1995). Lean women (BMI less than 19) had the lowest death rates of all, with death rates increasing steadily as BMI increased. A 2009 report published in the *Lancet* that examined data on close to 1 million adults found that blood pressure and blood cholesterol levels, both major risk factors for heart disease and stroke, increase linearly with increasing BMI starting from very low BMI levels (Whitlock et al., 2009). Although there was a higher mortality rate for the thinnest men and women, the excess was largely due to smoking-related diseases. Obviously some people have a low BMI because they are ill. They may have lung cancer, tuberculosis or perhaps an eating disorder such as anorexia nervosa. But for those who do not have a hidden infection or cancer and have good mental health, the evidence that a low BMI is biologically inappropriate is thin to say the least.

Mr and Mrs Average

If we measured the BMI of all the people in a particular country and plotted on a graph how many people there were at each BMI value we would find a wide spread of values. This spread of values is called a distribution. There would be some really skinny people with a very low BMI and some really fat people with a very high BMI, but most people would have a BMI close to the centre of the distribution.

So far we have focused on the ends of the BMI distribution. At the heavy end, we have seen that there are quite a lot of people

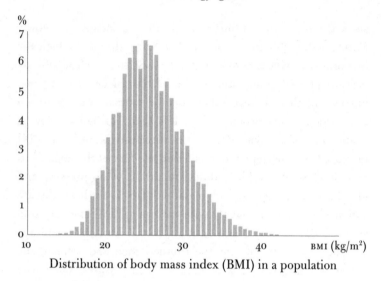

Distribution of body mass index (BMI) in a population

with a BMI over 30, far fewer with a BMI over 40 and very few huge people with a BMI over 50. The numbers also peter out at the light end of the BMI distribution. At either end of the distribution the number of people with very low or very high BMIs tails off, which is why the ends are often called the lower and upper tails.

The number of very thin people is decreasing and the number of very fat people is increasing, but what about the bulk of the population in between? The best measure of where the majority of people are on the BMI distribution is the average BMI. What's happening to Mr and Mrs Average? Unsurprisingly, they are also getting fatter. Between 1994 and 2004 the average male BMI in England increased from 26.0 to 27.3. This means that an Englishman of average height would have gained about 3 kilos in weight over the ten years. That's about half a stone. Mrs Average is also piling on the pounds. Between 1994 and 2004, the average female BMI rose from 25.8 to 26.9. It seems that everyone is getting fatter. The same trend is evident in the USA where the average BMI has risen from 25 in 1960 to 28 in 2002. Wherever you live, whatever your BMI today, it is likely that it would have been lower a decade ago for the same age and height. The media may have taken our eye off the ball. Focusing on the upper and lower tails of the BMI distribution, we have failed to notice that the whole population distribution of BMI is moving upwards. When it comes to putting on weight, there is no them and us. We are all in this together.

Chopstick thin

Dr Ayumi Naito completed her medical training at the Juntendo Medical School in Tokyo, and after practising paediatrics in Japan for several years moved to the UK. She now works in London in a clinic that provides health care for the 20,000 or so Japanese people living in the capital. Well used to study, she had no problem learning English. In fact, for Dr Naito, one of the most difficult things about living in London is buying clothes: 'I have to go shopping in France because I just cannot find my size in this country.' Had she been a fashion model rather than a doctor she would have been banned. Her BMI is just over seventeen.

Whereas one in four Japanese women has a BMI less than 20, the figure for the UK is one in fifteen women (7 per cent). Compared with Japan, the whole female BMI distribution in the UK is shifted upwards, dragging the skinny lower tail of the distribution with it. A BMI of 17 is not at all unusual in Tokyo. It is as rare as hens' teeth in London. Another interesting fact about Japanese women is that they don't appear to get heavier as they get older. British women get steadily fatter with increasing age.

17

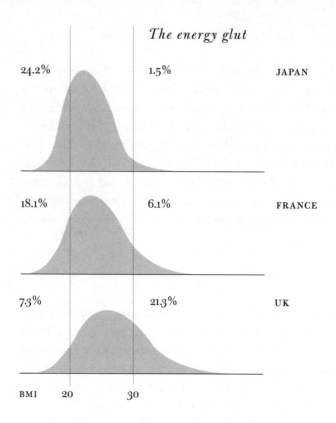

The energy glut

24.2%   1.5%   JAPAN

18.1%   6.1%   FRANCE

7.3%   21.3%   UK

BMI   20   30

Prevalence of obese and underweight females
in Japan, France and the UK, 2002 (WHO, 2010a)

The average BMI of a twenty-year-old British woman is 24, but
this rises to 28 by age fifty. It seems that living in Britain makes
you fat and the longer you live there the fatter you get.

For UK women in the upper tail of the BMI distribution there
is no shortage of 'plus size' clothing outlets. Making clothes for
this group makes sound business sense. After all, the whole
population is getting fatter. Sales can be expected to increase. Not
so at the thin end. Skinny women are on the decline in Britain.

18

Middle-aged and older Japanese women living in London have to compete with shopaholic twenty-year-olds for whatever slim wear is left in the shops. Shopping for clothes in Paris makes sound scientific sense. The population distribution of BMI in France is much closer to that of Japan. About 18 per cent of women in France have a BMI less than 28, and so in Paris it is still possible for a Japanese woman to find clothes that will fit.

So some populations are fatter than others. If you took a flight from Tokyo to Atlanta it would slap you across the face that there are few very fat people in Tokyo but truckloads of them in Atlanta. Perhaps the Japanese are just 'naturally slim'. Maybe the difference in BMI distributions between Japan and the USA is due to differences in genetic make-up. One way to check this out is to see what happens to the BMI of Japanese people who emigrate to live in the USA. They would have taken their genetic make-up with them. Do they stay slim or get fat? This question is what researchers at the University of Chicago set out to answer. They measured the BMIs of a random sample of Asian Americans living in the USA (Lauderdale and Rathouz, 2000). They found that US-born Asian Americans had a substantially higher average BMI and were much more likely to be obese than those born in Asia. What's more, for Asian Americans born outside the USA, their chances of being obese was directly related to the number of years they spent living in the USA. For example, compared with Asian American women who had lived in the USA for fifteen years or more, those who had lived in the USA for between five and fifteen years were half as likely to be obese, and the most recent arrivals, who had lived in the USA less than five years, were one-third as likely to be obese. Japanese people are not genetically slim. When they move to the USA they get fat like everyone else living there, and the longer they live there the fatter they get.

Most of the world's population is getting fatter

Increasing body fat is not a problem limited to a particular country. Wherever you happen to be living you can be reasonably sure that the population is getting fatter. Canada is not as fat as the USA but the Canadian BMI distribution is slowly sliding upwards. In 1978, the average BMI in Canada was 25. In 2004 it was 27. China has a relatively low average BMI but it has already started drifting upwards. Between 1989 and 2000, the average BMI of Chinese men increased from 21.3 to 22.4 and that of Chinese women increased from 21.8 to 22.4. It is the same story in Europe, Asia and South America. In 2006, the World Health Organization reported that there were 300 million obese adults living on planet earth and there will be many more in the future. Epidemics that affect the whole world are called pandemics. The USA may be the epicentre, but no country is immune.

There are of course health risks associated with being fat but these do not suddenly appear once your BMI crosses the dreaded 30 threshold. The risk increases steadily as BMI increases. Every five-unit rise in BMI is associated with a 30 per cent increase in the risk of death, with most of the excess coming from heart disease and stroke (Whitlock et al., 2009). When the British Medical Association lashed out against skinny models claiming that a low BMI was biologically inappropriate, it was making a value judgement rather than a scientific statement. Accumulating body fat is like accumulating debt. It is better to be $25 in debt than to be $30 in debt, but being only $20 in debt is better still. The reason that doctors like arbitrary thresholds is that they treat patients, and they need a way to decide who is a patient and who is not. If your BMI is 29 you do not (yet) have the disease 'obesity' but as soon as it reaches 30 you do. Obesity could be added to a growing list of medically defined diseases, along with high blood pressure and high cholesterol. According to the world-famous epidemiologist Geoffrey Rose 'There is no disease that you either have or don't have except perhaps sudden

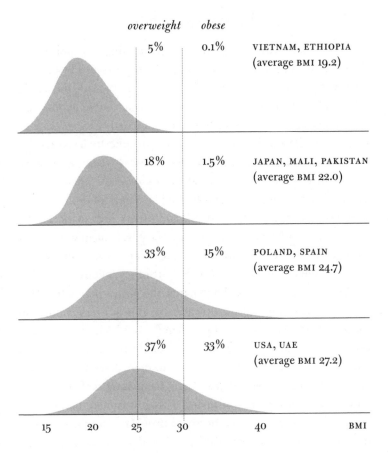

Male BMI distribution,
selection of countries, 2002 (WHO, 2010a)

death and rabies. All other diseases you either have a little or a lot of' (quoted in Smith, 2002). This is certainly true of BMI. As the average BMI increases, the average health risk we face increases with it.

The fat finger of blame

When it comes to tackling population fattening, our politicians appear to be on another planet. Being fat is portrayed by government as the just deserts of lazy gluttons who have made the wrong selections from a range of dietary and physical-activity choices. In January 2009, the UK government launched 'Change4Life', a major new 'lifestyle initiative' to stem rising obesity levels (NHS, 2009). According to the government, 'the way we live in modern society means a lot of us, especially our kids, have fallen into unhelpful habits.' Change4Life is backed by a pack of profit-hungry partners from the food and drink and fitness industries. Public health minister Dawn Primarolo said the government was 'trying to create a lifestyle revolution on a huge scale – something which no government has attempted before' (Seeney, 2009).

We have seen in this chapter that the entire population distribution of BMI is shifting upwards, in almost every country in the world. If this was a habit problem it would imply that almost everyone on earth had fallen into exactly the same unhelpful habits, a bad-habit pandemic on an unprecedented scale. This is of course nonsense. What is really happening is that the food and leisure industry is conspiring with government to convince the public that obesity is a personal problem, for which we can hold only ourselves responsible, rather than an environmental problem, for which governments and industry must take a large share of the blame. The personalization of fatness is arguably the most ambitious programme of mass propaganda since the war on terror. Decades of health research have shown that population fatness is an environmental problem and not a personal weakness. Our tendency to fall into unhelpful habits is the same as it ever was. As we will see in the next chapters, what has changed is that, thanks to motorization and mechanization, there are fewer opportunities to move our bodies than ever before, whilst at the same time we are besieged by a food industry that uses the best marketing brains in the world to sell us mountains of low-priced energy-dense food.

And the propaganda war is getting nasty. In June 2008, a spokesman for the UK National Obesity Forum called for the fattest children to be put into care. He was not alone. The same year, councils in England warned parents that obese children could be taken into care as an example of parental neglect. The councils were worried about the extra cost of educating fat children, for example having to buy larger chairs for schools. They had noticed that childhood obesity rates were shooting up and could see a budgetary crisis looming. The councils were clearly more worried about the children's effect on the school than about the school's effect on the children.

In the USA, Senator Joseph Carter proposed legislation that would require schools in Georgia to measure the BMIs of all their students and to post the school average on their websites. Perhaps naming and shaming would spur parents and teachers into taking steps to stem the fat tide. Health ministers in the UK had already ruled that parents should be notified if their child is obese so that they can make 'lifestyle changes'. Indeed, in 2008 the government changed the law to authorize the weighing and measuring of all UK children, with feedback to their parents on how fat their children are. Thousands of parents across the UK can now expect the results of the fat exam to fall though their letterbox. However, the good teacher it is, the government will also provide families with 'tips' on how to 'eat healthily and be physically active' so that failing families can do better next time.

We can predict which children will fail the fat test. It will be the same children who fail all the other tests. Fatness, like most health problems, is more severe among the most disadvantaged social groups (McLaren, 2007; Stamatakis et al., 2010). The inner London borough of Hackney is the most deprived in England and Hackney children are the fattest in the land. The link between fat and deprivation is also seen in the USA, where average BMI rises as educational level falls. The trend in BMI with educational level has made it easy to point the fat finger of blame. However, studies show that average BMI is increasing at the same rate across all educational levels (Truong and Sturm, 2005).

Most human attributes show a degree of variation within populations. Some people are taller than average whereas others are shorter. Some people are cheerful, whilst others are miserable. Human beings don't come off an assembly line. When it comes to BMI, some people have to be in the upper tail of the distribution, just like some people have to be in the lower tail. It is obvious that as the population average BMI increases and the whole distribution moves upwards, more people will cross the upper threshold that defines 'obesity'. Poor people may be more likely to be in the upper tail of the BMI distribution, but rich and poor are getting fatter together.

Black, white, tall, short, thin, fat, normal, obese: we live in a world of false dichotomies. This chapter has argued that the current preoccupation with obesity, which is simply the upper end of a continuous population BMI distribution that is slowly drifting upwards, is a pernicious form of propaganda that distracts us from coming to the obvious conclusion that the accumulation of body fat is a public issue and not a personal problem. Fatness is not a bad habit problem and it's not our genetic make-up. It's something to do with our environment. The next chapter explains what is really going on and why.

2

The motorization of movement

The previous chapter examined the population distribution of BMI and how over the years the whole distribution has moved upwards, a shift that inevitably means that more and more people enter the arbitrary categories of overweight and obese. I presented the view that fatness is an environmental problem rather than a personal failing, despite the best efforts of politicians and the food industry to convince us otherwise. This chapter and the next consider the environmental characteristics that are responsible for the upward slide in the BMI distribution. I would imagine that many readers will expect first and foremost a blistering critique of the food industry, which clearly provides us with more food energy than we need to maintain energy balance. However, although it is tempting to lay the blame for population fatness squarely at the tills of the supermarkets and fast-food chains, sadly this does not fit with the scientific evidence.

In 1995, Andrew Prentice, then head of the energy metabolism group at the MRC Clinical Nutrition Centre in Cambridge, carefully documented trends in energy intake and physical activity in the UK between 1960 and 1995, a period when the population was rapidly gaining weight (Prentice and Jebb, 1995). The best available data, corroborated from different sources, showed that average per capita food energy intake fell by 20 per cent over this period. This point deserves repeating. The prevalence of obesity was rising whilst average food energy intake was falling. The notion that fatness is all to do with greedy gluttons egged on by an insatiably profit-hungry food industry simply does stand up to scrutiny. Prentice found that the evidence implicating sloth to be far more compelling. Two major trends were highly correlated with increase in BMI: increasing car ownership and television viewing. In 1961, three out of ten households in Britain owned a car. By the mid-1990s, seven out of ten households had at least one car, and nearly a third had two or more. In the 1960s, the average Briton watched thirteen hours of television per week; by the 1990s this had risen to over twenty-six hours per week. Prentice concluded that the most likely explanation for increasing population fatness, despite a declining average energy intake, was that energy output was falling even faster than energy intake. We eat too much for our level of physical activity but our physical activity level is lower than ever before.

Scientific investigations into the causes of disease usually involve making comparisons of disease rates between people who are exposed to a potential causal factor and people who are not exposed. For example, by comparing disease rates in smokers and non-smokers we learned that smoking causes a whole range of different cancers. However, sometimes at any given point in time, the differences within a particular population in exposure to a potential cause of disease can be such that almost everyone is either exposed to the potential cause or else not exposed to it (Rose, 2001). In the context of fatness, Prentice found clear differences in exposure to potential causal factors by examining changes over time in one particular country. Another approach is to examine the differences between countries at a given point in time. A 2008 study from the University of Tennessee studied the relationship between active transportation, defined as the proportion of trips taken by walking, cycling or public transport, and obesity rates in Europe, North America and Australia (Bassett et al., 2008). Public transport was included with active transport because it usually involves some walking or cycling to reach the bus stop or train station. There was a strong and significant correlation between levels of active transportation and obesity rates. Countries with high levels of active transportation had low obesity rates. North Americans were the fattest and had the lowest levels of active transportation. The average distance walked by people in Europe was nearly three times higher than for people in the USA (382 km versus 140 km per year) and Europeans cycled nearly five times the distance (188 km versus 40 km).

Even more recent data has allowed us to examine the relationship between motor vehicle use and population fatness at a global level. The figure shows the relationship between the average BMI and motor vehicle gasoline consumption per capita for the 130 countries for which the necessary data were available. Each dot represents a country and the solid 'regression' line is chosen so that it comes as close to the points as possible. This line provides an indication of how average BMI varies as motor gasoline consumption increases. There is a clear strong relationship between average BMI and motor vehicle gasoline use. Where motor vehicle gasoline consumption is low, average BMI is also low, and where gasoline consumption is high, average BMI is also high. The

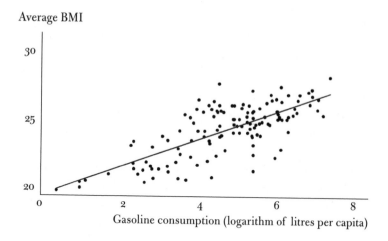

Male BMI and gasoline consumption in 130 countries
(WHO, 2010a; WRI, 2010)

association is present in both men and women, although strongest among men.

The reasons for the current overemphasis on food as the cause of fatness are both scientific and political. In the previous chapter I referred to the tendency of doctors to dichotomize people into healthy and the sick, when more often than not the first category merges imperceptibly into the second with no clear boundary. When it comes to fatness the doctors' dichotomy manifests as a preoccupation with obesity and its causes, rather than taking an interest in the average BMI of the whole population and trying to understand why it is rising. Our obsession with obesity is one of the reasons why we overlooked the effect on population fatness of the motorization of human movement. Whereas our fattening transportation infrastructure was built decades ago, and started our long plod towards population plumpness, we only woke up to the problem of fatness when large numbers of people started to crash through the obesity line.

We can imagine what effect a slow but steady upward shift in the whole BMI distribution would have on the prevalence of obesity. Because of the shape of the BMI distribution, with a peak in the middle around the average value, tapering out towards its upper end, initially there would be relatively small change in the prevalence of obesity, since only a small proportion of the population would be nudged over the 'obesity' threshold. However, as the average BMI gets closer and closer to the threshold (30 kg/m^2) the prevalence of obesity suddenly rockets. In other words, we would see a sudden large increase in the prevalence of obesity even if the average BMI was increasing at the same steady rate.

The current panic in the West about rapidly rising obesity levels is a consequence of forgetting that the whole BMI distribution was moving upwards. The slow but steady movement of the BMI distribution could be likened to the New York marathon, with the obesity threshold as the finishing line. Once the race starts, the whole population of runners starts moving towards the finish. For the first few hours no one crosses the finishing line but then the front runners start crossing in dribs and drabs. An hour or so later, the main lump of competitors reaches the line, and the number of finishers increases rapidly. In most wealthy countries the main lump of the population is now crossing the obesity threshold and so obesity is rising very rapidly. But there is little point searching frantically to find out what is happening now to explain the increase. The race towards obesity started decades ago. Here, however, the analogy ends because, unlike runners in a marathon, on our increasingly fat planet the runners just keep going.

The motorization of movement

Doctors and health scientists have been astonished by the sudden increase in obesity and have only recently started conducting studies looking for the causes. Most of these studies compare the food intake and daily activity levels of obese and non-obese people within a particular population at one point in time. The problem is that because we are now all exposed to the same transportation system, physical activity levels are at a record low across the entire population, and the factor that most differentiates the fat from the thin is food intake. By analogy, if everybody smoked and to exactly the same extent, and we compared people with and without lung cancer, smoking would not differentiate those with and without cancer. But this would not mean that smoking was not a cause of lung cancer. When it comes to fatness, scientists have been asking the wrong question (a focus on obesity rather than on average BMI), and have been using the wrong data (a focus on what is happening now rather than what has happened over the past half-century).

I will not go into detail here about the political reasons for the neglect of motorization as a cause of population fattening, other than to say that scientists tend to have a blind spot when it comes to gradual societal transitions and that they tend to limit their consideration of causes to those things that they believe are amenable to change. As we will see later, the notion that the de-motorization of society is either possible or desirable has effectively been barred from consciousness by a sustained bombardment of industry and government propaganda. But next I want to examine the impact of road danger on human movement. Road danger is a key environmental determinant of declining physical activity levels, even though it is scarcely given the time of day in contemporary public health literature. Entire populations do not passively give up walking and cycling because they are tempted to do so by the speed, comfort and glamour of motorized travel. On the contrary, they are driven off the street by deadly force, with a barrage of propaganda being required for its justification. The following case study illustrates what this means in day-to-day life.

An ordinary pedestrian death

Epidemiological research into the causes of disease or injury typically involves collecting information from those exposed and those unexposed to a particular potential causal factor, and then comparing the two groups to see what differentiates them. A disadvantage of this approach is that in order to glean just a small amount of reliable information on causal factors, the experiences of large numbers of people have to be distilled, and, in the sad alchemy of turning personal tragedies into risk factors, opportunities for a deeper understanding are sometimes missed. Whilst working in Auckland, New Zealand on an epidemiological study into the risk factors for child pedestrian injury, a sociologist with whom I was sharing an office challenged me to pick just one child road death and go into the case in detail. I learned more about the politics of road traffic injury from this case study than from years of data collection and number-crunching (Roberts and Coggan, 1994).

One of the children in the study was a ten-year-old girl who was walking home from school at 3.30 on a weekday when she was hit by a van. She was thrown high into the air, landing on the curb. An ambulance was called and when it arrived the paramedics attempted to resuscitate her. But it was too late. She was dead. A few minutes later a police officer arrived at the scene and filled out a traffic accident report. The report has boxes for the entry of information on the time, day, date and location of the accident, information on the vehicle, the driver and the road conditions. The speed limit on that particular stretch of road was 50 km/h. The questionnaire had a space for the officer's analysis of what happened and why. This is what this particular officer wrote:

WHAT HAPPENED

'Driver travelling east along two lane road – child stepped out onto roadway – into the path of the driver without looking, driver collided with child who was knocked into the air and landed on the grass verge.'

WHY THE ACCIDENT HAPPENED

Driver factors: 'Driver unable to stop in time due to sudden movements by child.'

Road factors: [nothing recorded].

Vehicle factors: [nothing recorded].

Other factors: 'Appears as though child has walked out onto the road without looking to her right.'

The report was later sent to the accident investigation division where the following accident summary was prepared:

'Child was standing on the footpath, without any warning she ran diagonally across the road into the path of an approaching van. The driver of the van was travelling towards [named] road and had no chance of stopping before she hit the child. Traffic safety branch have interviewed the driver and other witnesses and there is no indication of excess speed. Driver states that she was travelling about forty km/h. A scene examination by traffic seems to confirm this. Child was not on any medication at the time of the accident.'

The death was coded for statistical purposes as 'pedestrian: crossing road heedless of traffic, unattended child.'

One week after her death, as part of our research study we sent a civil engineer to the site where she died to measure the volume and speed of the traffic. Previous research had shown that the traffic characteristics at a particular site in the road network are remarkably constant from week to week. Vehicle speeds were normally distributed with an average speed of 58 km/h. Based on the speed data we calculated that the likelihood that any vehicle at that site would have been travelling at 40 km/h or slower, was less than 1 per cent. Some 86 per cent of vehicles at the site were travelling faster than the posted speed limit. Traffic flow was 877 vehicles per hour or 15 vehicles per minute. Assuming a steady traffic flow, the time available to cross the road would have been around four seconds.

Two months after the girl's death there was an inquest. The coroner began by introducing the court, pointing out that the 'coroner's court is not preoccupied with culpability; indeed it is not competent to decide on such matters.' He stated that the court 'aims to establish the facts', adding that 'this is not always easy'. After the police officer presented the case, the coroner asked 'did she have to cross the road to get home? I am interested to know why the road had to be crossed at that particular point.' The officer said he didn't know. The coroner then asked if the child had a hearing defect; none was reported. He pointed out that because of the widespread provision of road safety education in schools the sort of erratic behaviour shown by this child would be unusual in a ten-year-old. He said that children are repeatedly told not to jaywalk but that this child 'may have been doing a little bit of jaywalking'. He finally gave his verdict: 'I find that [name] died at [place] accidentally, sustained when she ran out into the path of an approaching vehicle without checking that the road was clear of traffic.'

This case study is not at all atypical. Mayer Hillman found that children are blamed for 90 per cent of child pedestrian deaths (Hillman et al., 2000). The *Oxford English Dictionary* defines an 'accident' as an event that is without apparent cause or that is unexpected. The use of the word 'accident' to describe child road deaths could not be more inappropriate. More is known about when, where and why child pedestrian–motor vehicle collisions occur than for almost any disease in childhood. It would make much more sense to talk about a case of accidental leukaemia. Calling road deaths accidents implies that no one is responsible and no one is to blame. The child either made a bad judgement or was just unlucky. But parents are responsible for their children and are not meant to put them in a position where poor judgement or bad luck can be fatal. Blaming a child victim really means that the parents are held responsible. Many bereaved parents live out their lives in silent desperation.

When the experiences of the hundreds of children killed and injured on Auckland's roads were aggregated, however, a very different picture emerged. Most of the children were injured close to home, often in the street where they lived. When we compared the traffic characteristics of the streets where injured children lived with a group of non-injured children selected from the general child population, we found that the main determinants of injury risk were the volume and speed of the traffic. The injury risk increased particularly steeply with rising traffic volume (Roberts et al., 1995). Children living in the busiest streets were fifteen times more likely to be injured than children living in the quietest streets. Traffic and not erratic jaywalking children is the cause of child pedestrian injury. Children get hit by cars because the cars are there. There is one street and it is either a place for children or a place for cars. Mixing the two, at least at average urban speeds, does not work without bloodshed. A vehicle driving down a residential street at 40 miles per hour packs more destructive energy than a bullet. If a child is unlucky enough to be hit, a single shot can kill and crossing a busy main road is like making a dash through machine-gun fire.

Epidemiological studies reveal associations. In this case, the association is between the volume of traffic and pedestrian injury risk. Whether this is a cause–effect relationship is a matter for judgement. It seems very likely, but perhaps we would like to confirm it with a natural experiment. What would happen to road deaths if the volume of traffic actually fell? Because the history of Western motorization is one of almost relentlessly increasing traffic, there has been little opportunity to answer this question. However, the Middle East oil crises of 1974 and 1979 provided a rare insight into what happens when traffic volume falls. Despite its clean green image, New Zealand is one of the most heavily motorized countries in the world and the 1974 energy crisis, which was accompanied by a fourfold increase in the price of petrol, hit hard. The government responded by introducing 'car-free days', when every car had to be off the road for one day each week. There was also a weekend ban on petrol sales, which lasted until August 1980. During the period of high petrol prices, child pedestrian death rates plummeted. Between 1975 and 1980 child pedestrian death rates fell by 46 per cent (Roberts et al., 1992). But when the oil started flowing and traffic volume resumed its upward trajectory, the number of children killed and injured on the roads increased along with it.

The oil price shocks of 1974 and 1979 also coincided with reductions in road death rates in the USA and Britain. The oil crises had revealed another layer in the Russian doll, another link in the chain of causation, that between the price of fossil fuels and the amount of danger on the roads. When petrol prices rise, fewer children die; when they fall, more children die. To a physicist this connection would seem obvious. Petroleum is chemical energy and the petrol in the fuel tank is the source of the energy that kills and injures. The tens of thousands of controlled explosions that drive the pistons that spin the wheels are fuelled by a steady stream of petroleum, and whatever obstructs this flow of chemical energy, whether it is war in the Middle East or hurricane damage to oil refineries in the Gulf of Mexico, there will be less road danger and fewer road deaths as a result. However, the oil crises were just a temporary blip and for the next thirty years traffic volumes would soar on the back of reliable supplies of cheap petroleum. Had nothing else changed, the bloodbath on the streets would have made current road death statistics seem trivial. But something else did change. The pedestrians and the cyclists got out of the way.

The pedestrians run for cover

A vehicle travelling at 40 miles per hour down a residential street is an obvious threat. Due to its mass and velocity, it contains enough kinetic energy to break bones and tear flesh. There is an equation in the physical sciences that tells us that the kinetic energy in a moving object is equal to one-half of its mass times its velocity squared. This means that a car weighing 1,500 kg driving at 40 miles per hour (18 metres per second) has $\frac{1}{2} \times 1,500 \times 18 \times 18 = 243,000$ joules of energy. A car moving twice as fast has four times as much kinetic energy. We do not need to have been hit by a car, or to have seen someone else being hit, to understand this. Evolution has hard-wired this knowledge into our brains. In the street, might is right. The person with the power is the person behind the wheel of the car. They hold the power to kill or disable and if they do kill or disable then they face little or no legal sanction. When faced by an assailant with a dangerous weapon you have two choices: to run or to fight. Picking a fight with raw kinetic energy is stupid and futile. The normal human response is to get out of harm's way and make sure that your children are out of harm's way too.

Child road deaths in Britain and the USA have been falling for decades. Death rates fell more steeply when oil prices were high, but the fact that the overall trend was down, despite rising road danger from increased motor vehicle traffic, meant that something else was going on. As the volume of road traffic increased and the streets became rivers of lethal kinetic energy, the pedestrians got out of the way. Parents kept their children indoors and those who could afford a car started driving rather than walking, even for short distances. The body counters at the ministries of transport of course claimed that death rates were falling because the traffic planners and police were doing a great job. Transport ministers proudly claimed responsibility for the fall in the number of deaths. No one bothered to count how many live people there were out on the streets.

There is another common response to an assailant with a dangerous weapon apart from getting out of the way. Obtain such a weapon for yourself. You would prefer that you and your children were safe inside a vehicle, rather than vulnerable on the outside. A survey of parents of primary-school children in North London found that most were very worried about road danger on the daily journey to school and that most would drive their children if they had access to a car (DiGuiseppi et al., 1998). And so begins the motorized arms race which drives the downward spiral of walking and cycling. If what I am saying is true, we would expect more pedestrian deaths among poor people who can't afford to use a car, and fewer pedestrian deaths among the wealthy people who can. Every death in England is recorded along with the occupation of the person who died. For dead children, the occupations of their mothers and fathers are recorded. When we looked at the data between 2001 and 2003 there were 205 children killed as pedestrians in road crashes in England and Wales. Of these, 71 children had parents who were 'long-term unemployed', 62 had parents who were in 'routine occupations', and 25 children had parents in 'managerial and professional occupations'. When taking into account the number of children in each group, the risk of road death for a child in the lowest social group was five times that of a child in the highest social class (Edwards et al., 2006). Walking surveys show that children from families without a car walk much more often than children in car-owning families (Roberts et al., 1997). Poor children are outside the car because they cannot afford to get inside.

Increasing road traffic has decimated walking and cycling in Britain, the USA, and most other highly motorized countries. Data from the UK National Travel Survey show that the average distance walked fell from 255 miles per person per year in 1975 to 192 miles in 2003. Over the same period, the average annual distance cycled fell from 51 miles per person to 34 miles. The distance walked by children has fallen by almost a quarter. Children today walk less than ever before in the history of humanity. It is likely that by 1975, when the UK Department of Transport first started measuring walking and cycling, most of the decline had already taken place. We can get an indication of the extent of the changes from old photographs and paintings. L.S. Lowry, the Manchester artist famous for his matchstick figures, painted Salford streetscapes in the 1920s and 1930s before the heyday of the motor car. His sombre skies show the pollution of industrial England, but his street scenes show a vitality that is completely absent today. The streets Lowry painted belonged to the people living in them. Now they belong to the car.

Traffic is not the only danger people consider when deciding whether to put on their walking shoes. For the past ten years, conducting clinical trials with trauma doctors on four continents, I have visited some of the most violent cities in the world. Many of these are in Latin America where the juxtaposition of conspicuous wealth and obscene poverty elevate routine urban violence to war-like proportions. There are forty homicides per 100,000 people each year in Medellín, Colombia, a murder rate over ten times higher than in London. When it comes to violence, even though it is people we fear, we feel most afraid when there are few people about. We feel safest in the peripheral vision of other people's awareness. We don't want to be stared at but we do want to be seen. In Medellín, as soon as I turn into an empty street my heart starts racing. In her book *The Death and Life of the Great American Cities* Jane Jacobs wrote how 'eyes on the street' help to keep the street safe (Jacobs, 1992). The more eyes the better, especially if those eyes can summon backup if there is cause for concern. She describes a brief urban drama in her neighbourhood. A man is seen dragging a young girl up a side street. The child is resisting, crying and shouting. A crowd quickly gathers. The eyes on the street did not like the look of what was going on. It turns out that the man is her father, the child is having a tantrum and that there is nothing amiss. The crowd disperses. On this occasion there was no cause for alarm, but people were concerned and came to help. According to Jacobs, this collective concern is what keeps communities safe.

Traffic takes eyes off the street. It divides the street. Interactions between people on the other side of a busy street are less likely to be noticed, voices might not be heard and the mood of interpersonal situations might not be understood. Yes there are eyes inside the cars, but when travelling at speed sight lines are polarized ahead, along the road, and not on the pavement. Would someone stop if someone was being attacked? Just as for road danger, the response to stranger danger is to get off the street and into a car, another vicious circle. Traffic makes a street seem hostile. This leads to more traffic and more hostility. The only people left on the street are the very poor people. Research shows that when considering whether to let their children walk to school, London parents fear stranger danger more than road danger. But are these parents worried that someone would drag their child screaming and yelling up a busy street or that they would be pulled into a passing car and whisked away? The latter seems far more likely. Enrique Peñalosa, the former mayor of Bogotá, Colombia, mused that children are like an indicator species for urban safety in the way that the presence of fish indicates whether or not a river is polluted. By taking children off the street, traffic increases our fear of violence. A colleague told me a story about a street party in England, where the street was closed to traffic for the day, as a demonstration project for a safe-street initiative he was involved in. The children from the street were outside playing. One of the residents said that it was an excellent event but asked my colleague, 'where did you get all the children from?'

So what effect do road danger and stranger danger have on our BMI? Anything that drives people off the streets and either into their homes or into their cars will reduce average energy expenditure (output) and will increase BMI. Energy output need

41

only be slightly lower than food energy consumption (input) for fat to accumulate. Take a 70 kg (11 stone) woman of stable weight, whose energy intake balances exactly her energy output. Part of her daily routine includes her walk to work, which is just 500 metres from her home. However, she is nervous about walking through some parts of her neighbourhood and saves up to buy a car. She parks her new car outside her home and from then on stops walking and starts driving. Unless she reduces her food intake to compensate for her decreased energy output, she will start to gain weight. In fact, one year down the track she will have gained about a kilogram of fat and will gain another by the end of the following year, and so on. Eventually, she will have to use more energy to move her heavier body and will regain energy balance albeit with several extra kilos. And by driving she will have made the streets just that little bit more dangerous and unpleasant for everyone else.

In this respect driving is like voting. Presidents do not fall on the basis of one single vote. Your vote becomes politically potent when aggregated across the whole of the electorate. Voting is a private contribution to a public mass action that carries the power to topple presidents. Driving to the shops does not make you a killer. The risk that you will kill anyone is minuscule. But the risk does exist and someone somewhere will kill a pedestrian while driving to the shops today. These small personal actions carry with them a tiny probability of causing harm, which when aggregated across the whole population have major public health implications. We will see later on how the motor industry and the car lobby attempt to personalize road danger. It is better for them that road death is seen as an errant act of a deviant driver or a 'jaywalking' child than the expected outcome of an unsafe system that kills 3,000 people every day, most of them pedestrians and cyclists.

Until recently public-health researchers had quantified the effect of traffic density on the risk of child pedestrian injury and the effect of childhood physical inactivity on BMI, but had not studied the direct effect of traffic density on childhood BMI. However, in a 2010 report in the journal *Preventive Medicine*, researchers from the University of California at Berkeley finally joined up the dots. Over 3,000 children between the ages of nine and ten years were followed until age eighteen. The researchers found a significant positive association between the density of traffic around the children's homes and their BMI levels at age eighteen (Jerrett et al., 2010). Even though the size of the effect was relatively modest, the fact that most children are exposed to high levels of residential traffic means that traffic could be responsible for a large proportion of childhood fatness. At least in the public-health literature, the ideology of blaming fat children and their parents is now coming under attack.

China takes to the roads

In the mid-1990s, after completing the Auckland study of child pedestrian injuries, I went to China, where the battle for the streets was well under way and pedestrians and cyclists were losing. By that time both motorization and fatness were well entrenched in Britain, Europe and North America, but for much of the world the road to fatness was still being built, and nowhere was this more evident than in China. At the Third Plenum of the Eleventh Party Congress in December 1978, Deng Xiaoping had set the gargantuan Chinese state moving in a new ideological direction. Economic reforms had been instituted to encourage private enterprise and promote economic growth. The Chinese economy started gathering momentum. Between 1980 and 1990, the gross national product per capita increased at a rate of 8 per cent per year. Industry, fuelled by an export boom, provided the driving force for economic growth. The immense internal market also contributed, with consumer goods such as cars, televisions and washing machines becoming available to the Chinese public.

During the 1980s there was a massive increase in motor vehicle use in China. Although by US standards levels of car ownership were low, there could be no mistaking the trend. Car registrations in China were growing at a rate of between 10 and 20 per cent a year and road freight was booming. Still, World Bank experts warned that shortages of transport would stifle the country's burgeoning economy and called for increased investment in road building. Until that time, mass transportation in China had meant bicycles. The urban road network was a web of wide, separated cycle lanes, which during rush hour became rivers of bobbing heads. At intersections, Chinese 'volunteers', garbed in orange uniforms, would blow whistles and wave flags to marshal the flow of bicycles. There were no parked cars but thousands of parked bikes, and more roadside cycle repair shops than there are petrol stations in the West. China, with tens of millions of bicycles, had the most equitable and environmentally friendly transport systems on the planet. But the era of the bicycle was over (Roberts, 1995).

I remember vividly my conversation with Professor Wang as we hurtled along the new dual carriageway between Chengdu and De-Yang City: 'De-Yang City, new city, used to be just a village', he told me. Before I could reply, we braked sharply as two children, probably no older than six or seven, made a dash across the road. I remember looking back, relieved to see that they made it across. About an hour or so later, at the health and anti-epidemic centre of De-Yang City, we compared statistics on child road deaths in China and the West. It was immediately obvious that road death and injury was a problem that East and West shared. However, the Chinese death rate of 9 road deaths per 100,000 children per year was already six times the rate in the UK (Wang et al., 1995). Chinese children were being taught a lesson. So far as the street was concerned, there had been a change in ownership. The streets were now places for motor vehicles, and motor vehicles are dangerous.

I knew that China was bracing itself for an enormous epidemic of road death and injury. What was not obvious to me at the time was what motorization would do for the Chinese waistline. I had not appreciated the links between car use, road danger, physical inactivity and fatness. I knew that road traffic injury was an energy problem. What I did not know at that time was that the normal human response to an environment riddled with deadly kinetic energy is to get out of the way, and that getting out of the way would lead to another energy problem, an explosion of fatness. By the end of the century, one-fifth of the 1 billion overweight and obese people in the world would be Chinese.

Although the prevalence of overweight (15 per cent) and obesity (3 per cent) in China is low by Western standards, the rate of increase is remarkable. Between 1985 and 2000, the prevalence of overweight and obesity in Chinese schoolchildren increased twenty-eight times (Wu, 2006). Of course, showing that two national characteristics are associated does not mean that one causes the other. There could be other differences between the countries that cause both to happen. For example, it could be that motor gasoline use is just a proxy for the overall level of economic development of a country, which could be the real cause of population fattening. But the associations observed at the country level are also seen in studies on individuals. Studies have found that people who regularly walk and cycle are less likely to be overweight or obese, and that acquiring a car significantly increases the chances of getting fat. A follow-up study of 2,485 Chinese men found that those who acquired a car were twice as likely to become obese and gain 1.8 kg more weight than those who did not get a car, even after controlling for food intake (Bell et al., 2002).

The runaway train of population weight gain

In the previous chapter we saw that the whole population distribution of BMI is shifting upwards all around the world. However, there is an important detail about this upward shift that I have not yet mentioned. As the distribution moves upwards the shape of the distribution changes (Swinburn and Egger, 2004). There is more movement in the upper tail than in the lower tail, so that the distribution becomes skewed. We are all gaining weight but the heaviest people are getting fatter the fastest. The tendency for weight gain to increase with increasing BMI has been likened to a runaway train that continues to gather speed (Swinburn and Egger, 2004). In an environment where there are few opportunities for human movement but that is overflowing with energy-dense food, the tracks run downhill. We have seen how road danger means less walking and cycling and a downward spiral in physical activity. This vicious circle interacts with other positive feedback loops at the personal level to guarantee accelerating weight gain. Although heavy people consume more energy when they move, they move less than lighter people. No surprise here. Even a short walk is a workout wearing a heavy body. With increasing body weight, back pain, arthritis and shortness of breath add to the discomfort of moving. Then low-self esteem from body dissatisfaction and the sense of personal failure when diets fail lead to comfort-eating, and, as we will see in the next chapter, there is no shortage of food to gorge on. And even when people are ensconced within their vehicles, the laws of physics lock the transportation system into a vicious circle of increasing fossil-fuel energy use. Large people require large fuel-hungry vehicles. Newton's laws predict that in a two-car smash the heavier comes out best. Evolution has also wired this into our brains. It is obvious that the bigger vehicle will come out best in a crash. The motorized arms race continues.

In this chapter we have seen how motor vehicle use contributes to population weight gain. Cheap petrol means more vehicles and more miles driven. The resulting road danger means higher injury risks for pedestrians and cyclists, which over time drives them off the streets into their homes and into their cars. Population energy output is slashed as petroleum replaces food as the primary source of energy for human movement. As the people get fatter, their addiction to fossil-fuel-based transportation gets stronger. We now face a new road danger. Motorized transport is 95 per cent dependent on oil, accounting for almost half of world oil use. Our fossil-fuel-powered transportation system is a threat to our health, and climate change is a threat to our survival as a species. It is no coincidence that the epicentre of the fatness pandemic is the USA, where per capita carbon dioxide emissions exceed those of any other nation. Cheap petroleum and wide open spaces conspired in creating America's low-density cities and car-dependent economy. Walking and cycling offer a lifeline to sustainability, but with one-third of US adults now obese, the prospects of a renaissance in walking and cycling seems remote.

3

Food and the petro-nutritional complex

We saw in the previous chapter how the motorization of movement and the rise in road danger have decimated population
average energy output all around the world. Historical data
from the UK, and more recent data from China, show that car
use greatly increases body fat accumulation. Human movement
that would previously have been fuelled by food energy is now
fuelled by petroleum. With millions of private passenger cars
out on the streets converting chemical energy (petroleum) into
kinetic energy (speed), our environment has become hostile and
hazardous and the normal human response is to keep well out of
the way. This leads to a vicious circle of increasing car use and
decreased walking and cycling that has serious consequences for
both health and climate change.

Motorization would have an even larger effect on population
BMI were it not for a compensatory mechanism. The normal
response to reduced movement is to eat less food (Blundell and
Gillett, 2001). Data from the China Health and Nutrition Survey

show that after acquiring a motor vehicle, average daily energy intake falls from 2,928 to 2,681 kcal in men and from 2,601 to 2,397 kcal in women (Bell et al., 2002). We are not therefore greedy gluttons. In fact, we are eating less food now than previously. However, those of us living in the paved parts of the planet now face a second environmental challenge that makes increasing population fatness inevitable. We are surrounded by an abundance of aggressively marketed, low-priced, energy-dense food. Attempts to diet are doomed to failure because they try to modify personal eating behaviours without changing the environment that shapes those behaviours. This chapter examines the food intake side of the energy balance equation. We will see that fatness is a normal response to a low-movement environment that is full of energy-dense food. We will also see that low-priced fossil-fuel energy is the reason why our food environment has become so fattening. But first we have to get to know our distant relatives.

Meet the relatives

Our evolutionary ancestors lived on the plains of East Africa about 3.6 million years ago. They were smaller and lighter than we are, had longer arms, smaller heads and a lot more body hair. They walked upright, although probably with a stoop, and would trek for miles every day foraging for food. Food was hard to come by. The sun beating down on the African savannah was the source of their food energy. Plants captured this energy in their green leaves and with water and nutrients from the soil built the carbohydrates that form the first link in the food chain. Insects, grubs and small animals would feed on the plants and large animals in turn would feed on these. Our ancestors' diet was mostly seeds and nuts and root vegetables but with an occasional

49

meat feast. They were scavengers more than hunters. They would compete with vultures and hyenas for the meat remaining on carcasses left behind after a big-cat kill. Foraging for plant foods and hunting and scavenging for meat was hard physical work. Fortunately, our ancestors managed to get their hands on just enough food energy to compensate for the energy they used finding it, and had enough spare at the end of the day for some serious reproduction.

Finding high-energy foods was not as easy as it is today. There was no food labelling. Some foods contained more energy than others and getting hold of this tucker could mean the difference between life and death. You are reading this book now because your relatives knew what they were looking for. Thanks to a run of genetic flukes and the law of natural selection our forebears were born with a sweet tooth which encouraged them to take advantage of the sweeter, more energy-dense plant species and the occasional cache of honey. They also developed a taste for fat, which, being an energy store, was an excellent source of high-calorie food. Because of their acquired taste for high-energy food, they had a survival advantage and flourished at the expense of the others. The ability to store energy as fat was another blessing bestowed by the genetic lottery. It gave them the edge during the lean times since they could call on their fat reserves when food was scarce. The irony of wasting the limited funds that are available for medical research searching for genetic causes of obesity is that your genetic make-up is perfectly adapted for your survival. The problem is that you were adapted to survive on a low-energy African savannah, and not in the energy banquet of the twenty-first century. But that is enough about our ancestors. How does modern man fill his stomach? To answer this question I did some food foraging of my own.

Undercover in the supermarket

I have just come back from an undercover operation in the supermarket. I wanted to find out how supermarkets influence the
amount of food that we buy. Posing as a shopper, I was making my
way up and down the aisles when, bam! There I was on holiday
in France. I could feel the summer sun on my face and the taste
of Bordeaux wine on my lips. I felt calm and happy. It was so
completely unexpected. I was just halfway down aisle seventeen,
having crossed from nappies and baby wipes into cheeses and
yoghurts. To begin with I was confused, and so I walked back
into babies and back again. Bam! There it was again. This time
I was in the garden of a house we had rented in Italy ten years
previously, enjoying a beautiful balmy evening, the summer sun
setting over the Tuscan hills. But then I recognized the smell
of warm bread and realized what they were up to. They were
messing with my limbic system.

The supermarket itself was a huge hanger of a building that
could have been a terminal at Heathrow Airport. The ceiling was
made of white perforated grille, through which you could see the
space-station silver tubes of the air-conditioning system. Here and
there the ceiling was perforated by a series of air vents and those
directly above the cheeses were wafting downwards the warm
aroma of freshly baked bread. Only by the cheeses, nowhere else;
a few paces back nothing, but two strides ahead and there was a
direct hit, and there I was, eating bread and brie on the banks
of the Garonne. Supermarkets don't actually make bread on the
premises; they bake bread using pre-prepared frozen dough.
They do this not to make money on bread but to make money on
other items through the strategic use of the sense of smell. Most
people think that baking smells are used to make shoppers feel

hungry. This is part of the plan, but through the strategic use of smell they are also tapping into the emotionally charged food memories lying dormant in the middle of shoppers' brains.

The limbic system is one of the deeper, more primitive parts of the central nervous system and is involved in many of our emotions, especially those related to survival such as fear and anger, but also pleasure, whether from eating or from sex. The limbic system is closely linked with long-term memory. By enabling our ancestors to differentiate between dangerous and pleasant situations, the limbic system helped to keep them alive during our perilous past. It has extensive neuronal connections to our sense of smell. Our primitive forebears may have scented food, opportunities for sex or dangerous situations well before they actually encountered them. These connections explain why scents and odours often reawaken distant memories and can evoke strong emotions.

From then on I walked the aisles with my eyes half-shut, surreptitiously scenting my way around the supermarket. Aisle sixteen, butter and margarine: there it was again with pin-point accuracy, the smell of warm bread. In the poultry section, another warm waft, this time of roasting chicken. I had come to examine the nutritional geography of the supermarket but now that I was exploring an emotional landscape I saw the store in a new light. Aisle twenty-three: barley waters and squashes, custards and tinned fruit, a quaint pick-and-mix sweet display and then chocolates. No smells this time but rich childhood associations with memories of being with my mother in the kitchen. Millions of years ago, our survival depended on our ability to scent the first whiff of danger, and the limbic system was part of an emotional guidance system that kept us safe. Modern humans have a larger forebrain and a greater capacity for rational thinking but our primitive brain structures are still intact. It is sad but true that people with some of the best forebrains in the world are now doing their very best to exploit our primitive brain structures in order to increase food industry profits.

Getting shoppers to buy more food is a mind game that the supermarkets know how to play. All of your senses are being targeted in order to get to your emotions. Shopping scientists are already using functional magnetic imaging techniques to find out how the brain's pleasure centres respond to different food brands (Fugate, 2007). Volunteers are put into a sophisticated brain scanner and shown various brands and packaging. If their pleasure centres light up in response to a particular brand, the food industry knows that they could be on to a winner. You can be sure that when they find the button to press, they will press it and you will buy. In 1954, researchers Olds and Milner inserted electrodes in the pleasure centres of rats' brains. They rigged up a system such that by pressing on a lever the rats could stimulate their own pleasure centres. When the rats were forced to choose between pressing the lever and eating and drinking, they chose pressing the lever and carried on pressing it until they died of hunger and exhaustion (Olds and Milner, 1954). In our case, our pleasure centre is linked to our mouth and stomach and so we keep eating long after it is good for us.

The rest of my undercover operation was uneventful. As expected, I found a massive free car park at street level with gently sloping travelators to carry customers from their cars to the store. Once past the bouncers at the door, you are into addiction alley, with the tobacco counter on one side and buns and cakes on the other. The tobacconist also caters for two other common social addictions, chocolate and lottery tickets. This section has its own tills, so that addicts can get their fix and get out without delay. Perhaps to distract shoppers from making the mental link between cigarettes and disease, there is a large flower stand right next to it. The rest of what shopping psychologists call the 'decompression zone' is organized around magazines, music and DVDs. The purpose of this area is not really to sell music but to slow us down so that we spend more time in the store, and the more time we take the more food we buy.

Once into the aisles you hit fruit and vegetables. Vegetables look best in the natural light by the door and getting something green in the trolley early on will assuage our health guilt and make it easier for us to reach for some seriously energy-packed food later. Here I can smell carrots, although this might be coming from the carrots themselves rather than the air vents. In the meat section the lights are brighter and everything is half-price or two for the price of one. On the corner of nearly every aisle (known as en-caps in the trade) there are massive stands of 2-litre Coca-Cola bottles, just in case we forget about them.

I notice that the aisle that I am walking down is called Market Street. It is a wide brightly lit aisle, temperature-controlled, and there is music playing. Of course there is no road danger here. It is very obvious that the supermarket is trying to re-create inside of the store what has been lost on the outside. They have built in this private space what we might wish for in the public space – a safe and pleasant shopping environment that looks and feels like a farmers' market. Milk, bread and other staples are deep inside the store, ensuring that shoppers spend time searching for them, increasing the chances that they will make other unplanned purchases on the way, but all of this is standard supermarket psychology. I had not intended to buy anything, but left with milk, a jar of coffee, two carrots, three onions, a box of cereal, a tin of custard powder and a loaf. Explain that if you can. The supermarkets know exactly what they are doing and they are all made to sell us energy and they all look just the same.

Supermarkets and the petro-nutritional complex

Population fatness depends primarily on the transportation system and the food system, since these determine our average energy output from movement and our average energy input from eating. Although this chapter is about food, in reality transport and food are intimately interconnected and nowhere is this more apparent than in supermarkets. Fatness today depends critically on the sale of two consumer goods, petrol and food, and the supermarkets are a major player in the sale of both. When it comes to food, supermarkets have a monopoly. British shoppers spend 93 pence of every pound spent on food in one of the supermarket chains. As for petrol, the supermarket share of retail sales is rising rapidly. In 2007, the supermarkets' share of petrol sales was 41 per cent of the total market. The reason that supermarkets can undercut independent petrol retailers and put them out of business is that even if their profit margin on petrol is low, they can make more money from selling more food. Supermarkets sell energy, either as food or as petrol. From a business perspective this is a shrewd development. We have seen that as people move less, all other things being equal, they tend to eat less. Over the long term, increasing motor vehicle use could undercut food sales. One way to avoid this is to control both sides of the energy market. The link between petrol and food sales is so strong that we could think of supermarkets as simply the retail arm of a petro-nutritional complex. Supermarkets don't care how they make profits, whether by selling us petrol or food, but either way it makes us fat.

It also makes good business sense for supermarkets to sell fuel for cars and people in the same place. Discounted petrol and free parking encourage shoppers to take their cars to the supermarket, which means that they will buy more food than they would if they had to carry it home on foot, bicycle or by public transport. Currently, over three-quarters of British households do their main food shopping by car (DOT, 2007). Shopping by car also makes us fat because we use up less bodily (food) energy whilst doing our shopping. Finding (foraging for) food and bringing it home is one of the main human activities and always has been, and shopping by car has slashed personal energy output.

Making us fat might not be the intention of the supermarkets but it will boost their profits. When it comes to food energy consumption, wearing a heavy body is like driving around in a gas-guzzler. A heavy person must consume more food energy just to maintain a constant body weight. There is simply more metabolically active flesh to feed. Similarly, fat populations consume considerably more food than lean populations. It has been estimated that the amount of food energy consumed by a 'fat' population (average BMI 29) is around 20 per cent more than that consumed by a 'normal' weight population (average BMI 25). (Edwards and Roberts, 2009). The heavier our bodies become, the harder and more unpleasant it is to move about in them and the more dependent we become on our cars. The more food we eat, the more petrol we consume; the more petrol we consume, the fatter we get; the fatter we get, the more food we eat – a frighteningly fattening vicious circle.

Shopping by car may be good for supermarkets, but it is bad for road safety. Shopping accounts for 20 per cent of all trips in Britain, and if these trips are made by car, then our roads will be considerably more dangerous for people who want to get their physical activity moving about their towns and cities, rather than moving about in the gym (DOT, 2007). It is worth pointing out that the big-name supermarkets are just one component of the petro-nutritional complex. Whilst supermarkets have become petrol stations, petrol stations have become mini-supermarkets and most sell a range of foods, beverages and energy-dense snacks. Nowadays, humans and cars graze on the same pastures.

Food glorious food and everywhere

Thanks to motorization and mechanization there has been a proliferation of machines that reduce personal energy use, from escalators and moving pavements to machines that blow the leaves off the path. The amount of food energy that we need to consume to balance our reduced energy output is less than at any other time in history. The only way to stem the fat tide is to consume less petrol and less food. But this is health advice that the car industry and the food industry find hard to stomach, and so the guidance we receive from government is that we need to eat more healthy food and to do our movement in the gym. How the correct dietary message to *eat less food* has been morphed by the food industry into a message to *eat more (healthy) food* is a propaganda feat worthy of any Stalinist dictator and has been marvellously documented by Marion Nestle in her book *Food Politics: How the Food Industry Influences Nutrition and Health* (Nestle, 2003). Eating food that has a high nutritional content is important for health, but it is the quantity of food energy consumed that makes us fat. We must not forget that fatness rather than vitamin deficiency is the pandemic that the Western world is facing.

So how much food energy do we need? The energy requirement of the average sedentary North American is now around

2,000 kcal per day. The US food industry produces enough food to provide around 3,900 kcal per person per day, almost twice the population's energy needs (Ludwig and Nestle, 2008). There is certainly no shortage of food energy on the shelves. The two ways that food companies can increase profits is by encouraging us to eat more and by selling us high-profit-margin processed foods, which also tend to be high in fat and refined sugars – in other words foods that contain a lot more energy. Both strategies make us fat. It is no accident that food marketing encourages us to eat larger portions, with more snacking between meals and greater consumption of sweets and soft drinks. This is where the profits come from.

The laws of economics indicate that when prices fall consumption increases. Although there have been short-term food price increases, the overall trend is down. Food has never been cheaper. Between 1980 and 2000 food prices in the USA rose by 3.4 per cent per year, which is slower than the 3.8 per cent average rise in the inflation rate. This means that the relative price of food has fallen over time. In the UK, the percentage of household income spent on food has fallen from around 20 per cent in 1970 to 10 per cent in 2002 (Lobstein and Jackson-Leach, 2007). Added to this, the price of the most energy-dense foods, the fats and sugars, has fallen more rapidly than the price of the least energy-dense foods, the vegetables, fruits and grains. The price of sugary drinks has fallen the fastest, and as prices have gone down consumption has skyrocketed. Soft drink consumption in the USA has increased by 300 per cent in the past twenty years (Harrington, 2008). Currently, more than three-quarters of US schoolchildren drink at least one soft drink per day. This is not surprising considering their low cost and the fact that US children watch more than ten food industry adverts every hour spent in front of the television, most of which are for soft drinks and sweet snacks (Story and French, 2004). North American children are drowning in liquid calories. And the rest of the world is jumping in to join them. Coca-Cola boasts 1.5 billion consumer servings per day. In summary, price signals to shoppers strongly encourage a change in purchasing patterns to the cheaper high-energy foods.

Although the causes of population fatness are easy to understand, some doctors have managed to muddle the message. News headlines about research into the genetic causes of obesity are common and leave the public with the impression that the accumulation of body fat is something to do with their genes. I explained in the previous chapter that the scientific search for the causes of disease or illness depends on the presence of variation. Because we are all exposed to a car-based transportation system, food intake is the variable that explains within population differences in fatness. However, we are now reaching a stage where, at least in Europe and North America, we are all exposed to the same food environment. Once this happens, the variable that explains within population differences in fatness is genetic differences. Genetic differences do not explain why populations everywhere are getting fatter, although they may explain some of the variability within populations at one point in time. Poor farm workers in Peru who move to the city looking for a better life take with them the same set of genes. They get fat because in the city they move less and eat more. Genetic factors might make a small contribution to explaining why some of them get fatter faster than others, but they do not explain why the population average BMI is rising in nearly every country in the world, which is the important question.

The preoccupation with genetic factors largely reflects the commercial interests of drug companies, which stand to make fat profits from the drug treatment of people in the upper tail of the population weight distribution. Identification of a genetic predisposition to fat accumulation would give them an important sales opportunity. However, research into the genetic causes of obesity is expensive and rather low yield, and is the sort of research that the pharmaceuticals industry would prefer to be done with public funds in the universities, although they will no doubt keep a close eye on what might be turned into private profits. Somehow or other they manage to convince public funding bodies that this is what 'real science' is about. The search for fat genes will no doubt continue until the drug companies find something that they can make money from.

You don't need a degree in medical genetics to understand why the world is getting fatter. In the context of reduced energy output, the combination of biological hard-wiring and food abundance, particularly when the marketing might of the food industry is added to the mix, leads to energy intakes that regularly exceed energy output. But where does all the cheap food energy come from? How did our planet move from famine to feast? The next section explains how the fossil fuel sucked out of the earth's crust finds its way into our adipocytes. It may sound implausible but it's all energy in one form or another.

Fossil fuel and food

The energy story begins with sunlight. The sun has always been the primary source of food energy. The earth, in its daily communion with the sun, is blessed with about 200 watts of solar energy per square metre. Plants turn their leaves to receive this energy and, after taking in carbon dioxide from the atmosphere, through the process of photosynthesis, make the food that sustains life. All our ancestors had to do was to find it. However, foraging for food, although easy in the good times, was unpredictable. Short-term climatic changes caused devastating famines. Early farming was harder work but the food supply was more reliable and any given amount of land could support more people. Using animals to pull ploughs, the use of irrigation and animal manure as a fertilizer increased the productivity of the land, but droughts, floods and natural disasters still caused crop failures and famines. For centuries, the amount of food energy that could be harvested from a given amount of land was severely constrained. But then something special happened. Practically limitless supplies of cheap oil changed everything. The photosynthetic equation taught in school is carbon dioxide + water + sunlight = food + oxygen. However, in the presence of fossil fuels, the same set of ingredients = lots of food + oxygen. From the 1940s onwards, fossil-fuel-based fertilizers, pesticides, irrigation and mechanization massively increased food yields. Remarkably, a slimy back sludge that gushed from the ground had started a green revolution. As the physicist Albert Bartlett observed, agriculture became the use of land to turn fossil fuel into food.

Breaking the nitrogen barrier was a major milestone. Plants need nitrogen to be present in soil in order to grow. Although there is plenty of nitrogen in the air, low levels of nitrogen in the soil limits crop yields. The Haber–Bosch process is a chemical reaction that turns nitrogen and hydrogen into ammonia that can be used in farming to increase soil nitrogen. Its discovery marked a turning point in modern agriculture. The world is now critically dependent on the Haber–Bosch reaction to produce the estimated 100 million tonnes of nitrogen fertilizer that help to grow the food that feeds a third of the global population. The reaction is highly energy-intensive, so much so that the Haber reaction consumes almost 2 per cent of the world's annual energy supply. Thanks to an eightyfold increase in the energy inputs to farming, the twentieth century saw a fourfold increase in average crop yields. It is true that millions of people in the world remain undernourished, but this is not due to a global shortage of food.

Growing food is just the first stage in the food supply chain. Fossil fuels also power the processing, packaging, distribution, storage, retail and cooking of food, as well as the disposal of food waste (Friel et al., 2009). We have already seen that the food and transportation systems are inextricably interwoven. The rise of the supermarkets went hand in hand with increasing car ownership and motorway building. Investment in road building allowed out-of-town supermarkets to flourish since shopping by car allows shoppers to travel further and to buy larger quantities of food. This trend is reflected in UK shopping patterns, which show that while the number of shopping trips has declined over the past twenty years, the number made by car and the average distance travelled increased. Selling larger quantities of food meant larger profits for the supermarkets. Because the best value for money in food shopping was at supermarkets only accessible by car, this increased the demand for car travel.

The idea that fossil fuels are a key ingredient in our food is not an easy one to swallow. Nevertheless, the sharp rise in food prices that accompanied the sudden oil price hikes of 2008 showed clearly that cheap food depends on cheap oil (IFAD, 2008). In July of 2008, oil prices hit an all time high and the politics of oil and food moved to centre stage. Although at the time, much of the media debate centred on biofuels, their contribution to rising food prices was a relatively minor one, the main driver of food prices being sky-high oil prices. The resulting food price hikes led to riots in Bolivia, Burkina Faso, Cameroon, Egypt,

Indonesia, Ivory Coast, Mauritania, Mozambique and Senegal. Petrol tanks and stomachs were in competition long before the wealthy nations introduced policies on biofuels.

Oil is energy and food is energy, and since the start of the agricultural revolution the price and availability of oil have determined the price and availability of food (Baffes, 2007). Because the world has enjoyed almost a century of cheap oil, disregarding its environmental consequences, there has never been so much food energy in our environment. However, like a drunkard the morning after an oil binge, the world has woken up to the headache of climate change. The scientific evidence that our burning of fossil fuels is affecting the most fundamental determinants of human health (food, air, water) is overwhelming. The effects of climate change will be widespread; some countries will come off worse than others but no country will be spared. The expected health risks include effects on the availability of food and drinking water, especially in Africa, changes in the geographic spread and rate of transmission of infectious diseases, extreme weather events, mass starvation, forced migration and violence. It is of note that the US and UK militaries are increasingly concerned by the 'security implications' of climate change, or, perhaps to put it more accurately, by the security implications of millions of desperate climate refugees fleeing a starving continent. What can be done to stop this happening will be explained later, but now let's take a closer look at the link between food and climate.

Food and greenhouse gas emissions

Because of its dependence on fossil fuels, the food system is a major contributor to global warming, although due to the interweaving of the food system with the transport system it is

63

difficult to estimate how much the food system alone contributes to greenhouse gas emissions. The Intergovernmental Panel on Climate Change (IPCC) estimates that agriculture is responsible for 10–12 per cent of global greenhouse gas emissions (Friel et al., 2009). However, this figure does not include carbon dioxide emissions from deforestation, and to produce our food we cut down a lot of trees. Growing trees absorb carbon dioxide from the atmosphere and lock it away where it cannot contribute to global warming. When forests are burned or cut down to clear the land for farming, the stored carbon is released. When the effects of agriculture on deforestation are factored in, it is estimated that agriculture is responsible for between 17 and 32 per cent of greenhouse gas emissions. An EU report estimated that the food supply chain is responsible for 31 per cent of total emissions in Europe (EU, 2006).

When it comes to the greenhouse gas impacts of specific food types, we see that meat is heat (Friel et al., 2009). One-third of the world's land surface is used for livestock production, either to provide pasture for grazing or to grow grain for cattle feed. Meat production causes large-scale deforestation, and the production of grain for cattle feed requires the use of energy-intensive fertilizers. In Brazil, land clearance to grow soya for cattle feed is causing rapid and rampant deforestation. Methane released from animal manure and from enteric fermentation is a particularly powerful greenhouse gas. Although methane released from industrial live-stock rearing accounts for a large part of agriculture's contribution to global warming, such intensive production methods would not have been possible without oil.

Our fossil-fuel-powered food system is contributing to both population fat accumulation and climate change. Climate change and weight gain are different manifestations of the same planetary malaise: our dangerous relationship with fossil-fuel energy. Future generations will pay for the damage we are doing to the global climate, but we will pay for our body fat accumulation. Those who can afford it will pay to use tedious exercise machines in sweaty gyms. They will pay for drugs that stop their bodies from absorbing fat so that it drips out of their bottoms like slime. They will pay for liposuction and gastric surgery, diabetes and heart disease.

Fatness is an environmental problem and not a personal shortcoming. Dieting does not work because eating and movement are governed by the food and transportation environments, and diets leave both of these unchanged. There are some seriously clever people out there whose job it is to sell us cars, petrol and food whether we want them or not, and they know more about how our brains work than we do. If we want to regain control over the shape of our bodies and prevent catastrophic climate change, we must take back control of the environment and dismantle the petro-nutritional complex.

4

Money

It is 8.15 in the morning and the sixth floor of the Bloomberg
Tower on Manhattan's lustrous Lexington Avenue is buzzing
like a beehive. With each ting of the elevator, a small cloud of
worker bees is ejected and disperses, but only to gather again
as the morning swarm settles around the long grey honeycomb
of the complimentary breakfast bar, its square open cupboards
packed with boxes of cereal, granola bars, cookies and crisps.
As they gulp their coffee and guzzle their snacks, the office
staff gaze upwards at the markets data and business news that
are streaming past them on the brightly coloured LED screens:
DOW down, FTSE 2004.68, banks shed toxic assets, ICAP
down 21.5, car makers seek bailout. I am not sure what these
numbers mean for the health of the global economy. I never
learned to speak the economic Esperanto of stocks and shares.
What I do know is that money and profit are crucial links in the
chain of causation that leads from fossil fuel energy to fatness
and climate change.

I have been summoned to New York, along with other road-safety researchers from around the world, to advise billionaire philanthropist and New York mayor Michael Bloomberg on how best he can use some of his fortune to improve road safety in poor countries. The first person I meet is the director general of the FIA Foundation. The Foundation was established in 2001, by the Federation Internationale de l'Automobile, which is the governing body for motor sport and the federation of the world's leading motoring organizations. Its claim to road safety expertise appears to be the relatively low death toll among Formula One racing drivers despite their heroically high speeds. We had exchanged opinions about the appropriateness of FIA involvement in global road safety in the comment section of the *Guardian*. My article had pointed out that one important difference between the grand-prix racing circuit and the roads in poor countries is that there are no pedestrians, cyclists, or children on the racing track. (Roberts, 2007).

The Foundation wants more international development aid to be 'invested in road infrastructure schemes in the developing world to prevent "killer roads"'. But should road building be considered to be development at all? Would any aid money be better spent on programmes that really improve human development, in areas like education, health and family planning? If money has to be spent on transport, then considering that most of the victims of road traffic injury in poor countries are pedestrians and cyclists, surely the money should be spent promoting safe walking and cycling. Such questions are unlikely to be resolved in the five minutes before the meeting begins, although I suspect that the mayor may be more receptive to them. The day before, he had launched a report by the New York City Panel on Climate Change that predicated that water levels around New York City would rise by 2 foot in the coming decades, with temperatures rising between 4.0 and 7.5 degrees (NYCPCC, 2009).

This chapter examines the nexus of roads, oil and cars and the central position that these have in relation to corporate wealth. We have seen how rapidly increasing motor vehicle use and the road danger that it causes has slashed human physical activity, contributing importantly to population fatness, whilst the carbon dioxide emissions caused by motor vehicle transport are worsening climate change. But why are the world's roads so perilous? Who benefits from all this dangerous traffic? Simply knowing that road transportation is damaging personal and planetary health will not help us to save ourselves and our world unless we understand why the world has been designed for motor vehicles in the first place.

The main reason why car travel is annihilating human movement and putting our planet in peril is that motor vehicle travel is highly subsidized. Motorized transportation depends on three essentials: roads, oil and vehicles. Although most of the world's population will never own a car, road building is invariably funded by public funds, in rich and poor countries alike. Road transportation is 95 per cent oil-dependent, and ensuring a steady supply of cheap oil also involves massive public expenditures. Then there is the automobile industry, which received billions of dollars of taxpayers' money at a time when thousands of small businesses were going to the wall. In December 2008, US car makers went cap in hand to Congress seeking a $34 billion bailout package and received the best part of it.

And it does not stop there. Motorized transport causes a mountain of suffering. Who pays for the road traffic crashes that kill 1,000 children per day and permanently disable ten times as many? Who pays for transport-related air pollution and the cardiac and respiratory diseases it leads to? Who pays for physical inactivity and the obesity, diabetes, heart disease, stroke and cancer it causes? And who will pay for climate change? These are the real social and environmental costs of motorized transport, but it is the public and the global environment, not the people who benefit from motor vehicle use, who pay. The table below shows the top ten corporations in the world according to the Fortune 500 annual ranking for 2008. Eight of the top ten are oil companies or car makers. The corporations that sell oil and cars pack enormous economic and political clout, with revenues higher than the gross domestic product of many developed countries. This chapter argues that we drive around in cars because the most powerful corporations in the world prefer it that way.

Fortune 500 top ten ranking for 2008

RANK	COMPANY	SECTOR	REVENUE ($m)	PROFIT ($m)
1	Wal-Mart Stores	Supermarket	378,799	12,731
2	Exxon Mobil	Oil	372,824	40,610
3	Royal Dutch Shell	Oil	355,782	31,331
4	BP	Oil	291,438	20,845
5	Toyota Motor	Cars	230,201	15,042
6	Chevron	Oil	210,783	18,688
7	ING Group	Finance	201,516	12,649
8	Total	Oil	187,280	18,042
9	General Motors	Cars	182,347	-38,732
10	ConocoPhillips	Oil	178,558	11,891

Africa needs more roads

Dr O looks tired. He was up all night. As a trauma surgeon at a major trauma unit in Nigeria, he is on call most of the time and he is often called. 'Last night I admitted a woman, twenty-two weeks pregnant, hit by a truck. She's brain dead but the baby inside her is still alive', he explains. If the baby is delivered now it will die. Its lungs would be too immature. But keeping the mother's body alive will not be easy in Nigeria. Despite being the largest oil producer in Africa with 36 billion barrels of proven oil reserves, the country's electricity supply is unreliable and there are frequent power cuts, even in the hospitals.

What is not in short supply in Nigeria are trauma patients. On a previous visit to Nigeria I saw a body by the side of the road; a headless corpse, most probably a recent victim of a high-speed road crash. Although the roadsides are crowded with pedestrians, few roads have pavements or safe places for pedestrians to cross, and at night there is hardly any street lighting. And in the oil-rich Niger Delta violence is endemic. In 2006, the Movement for the Emancipation of the Niger Delta (MEND), an armed faction dedicated to ending exploitation and environmental degradation in the region, warned foreign oil companies to 'leave our land or die in it'. In 1995, Nigerian poet-turned-activist Ken Saro-Wiwa had used non-violent protest to draw attention to the injustices in the delta and was executed by the government for his efforts. MEND was trying a different strategy.

The wealthy world needs Africa to build more roads so that Western car makers can remain profitable. The market for cars in high-income countries is nearly saturated. In the year 2000, there were 769 cars per 1,000 people in the USA and 441 per 1,000 people in UK. Although there is some turnover, as old and damaged vehicles are taken out of stock, a process that was recently given a boost with a generous injection of public funding, the main prospect for a growth in sales is in Africa and Asia. Nigeria has 11 cars per 1,000 people and India has 7 per 1,000 (GZT, 1999). To survive, the car industry must sell more cars, and to make sure that it can poor countries must build the roads to accommodate them.

Getting impoverished countries to spend public money on road building requires some serious propaganda. Poor countries have lots of pressing problems to deal with. The most pervasive misinformation is that road building is good for development. In 2005, the British prime minister Tony Blair launched the Report of the Commission for Africa. The Commission's objective was to diagnose African woes and make a prescription for a better future. Its conclusion was that Africa needed more roads. More important than health care, AIDS prevention, security or better governance, road building it was argued would jump-start the stalled economy of a continent mired in misery for decades.

The Commission's analysis was simple. Africa is poor because its economy is not growing. Improving its transport infrastructure would make its goods cheaper, allowing Africans to break into world markets and trade their way out of poverty. Of the estimated \$75 billion needed to implement the Commission's recommendations, 27 per cent would be spent on infrastructure, mainly road building, with 13 per cent spent on AIDS, and 10 per cent on education.

If reducing the costs of getting African goods to Western markets is really the cause of African poverty, as Tony Blair and the Commission for Africa claim, Britain could help Africa's poor by reducing the transportation costs for African goods once they reach Britain. Like many wealthy countries, Britain has high levels of fuel taxation. In most of Africa, fuel is not taxed but subsidized. In 2004, a litre of super gasoline in Nigeria retailed for US$ 0.40. The corresponding cost in the UK was US$ 1.56. British politicians would not contemplate reducing fuel taxes since these are such an important source of government revenue. Indeed, the money raised from fuel sales in Britain helps to pay for the huge state apparatus required to service a car-based transportation system. This includes the police needed to enforce road safety laws, a judiciary, a system of pre-hospital and hospital care, and a social safety net for injured victims. African fuel prices do not even cover the costs of road maintenance. According to the World Health Organization, the economic loss associated with road traffic injuries in poor countries is around 2 per cent of GDP, nearly US$100 billion, twice the amount they receive in development aid (Peden et al., 2004). Reducing the cost of road transportation in Africa might be good for trade, but it is not so good for most Africans.

In 2006, the UK Department for International Development commissioned transport expert Professor David Banister at University College London to collate the scientific evidence on the link between road building and development (Banister et al., 2005). Although he did find a statistical link between the road infrastructure and the size of the economy, he could find no evidence that the former caused the latter. It is no surprise that rich countries have more roads than poor countries. Wealthy countries have more cars and so there is a higher demand for roads. There are more swimming pools in wealthy countries but no one would claim that swimming pools are 'central to development', which is what the World Bank claims for road building. However, the report did point out that the congestion resulting from rapid motorization hampers economic productivity and that the poor bear the lion's share of the negative impacts of road transport. At the Department for International Development the report was given a hasty burial.

Making profits in Africa

In the summer of 2004, I was asked to give a lunchtime talk on road safety to managers at a UK-based investment company. The company's mission is to create wealth in emerging markets, particularly in poor countries, by investing in 'sustainable' private-sector businesses. In 2003, the investment company made a pre-tax profit of £45 million ($85 million), some £15.6 million of which was made from its investments in Africa. The only problem was that its business activities in Africa left behind a trail of road deaths and injuries.

Before giving my talk, I had asked for some background statistics on the traffic injuries associated with the company's activities. I was sent a summary of the fatal 'accidents' attributable to the businesses in 2003. In that year, there were thirteen fatal injuries 'directly attributable to the work activities of the company', about half of which were transport-related. A young Tanzanian girl had died after being run over by a company-managed truck. She did not die at the roadside, but 'following poor care she died in the hospital 3 days later'. Another two-year-old Tanzanian child was crushed to death by a tractor, and in Swaziland a contractor's truck struck and killed a child on his way to school. These were the deaths that the company knew about. Road deaths are notoriously under-reported in poor countries (Roberts, 2005). The actual numbers of road deaths can be four times higher than the number recorded in police statistics, and the number of serious injuries can be seventy-five times higher. And deaths are just the tip of the injury iceberg. For every death there are about fifteen injuries requiring hospitalization and seventy minor injuries. You might think that child deaths would have been taken seriously by the company, but they were not even mentioned in the Annual Report. The Report said that the businesses in which the company invests must observe 'minimum standards in relation to health, safety and social issues'. Clearly, for some children the minimum standards were not enough.

I asked the company for more information about the deaths and injuries associated with its activities. I asked if they keep records on non-fatal injuries, whether there was a proper police investigation in each case, whether the families were compensated and if so by how much, and what measures had been put in place to prevent future deaths and injuries. I received no reply and so I wrote to the people who own the company at the Department for International Development (DFID). Curiously, the company in question is wholly owned by the UK Government. Hilary Benn, then secretary of state for international development, explained that his department 'does not get involved in its day to day operations' and does not hold 'detailed data of this sort'. He passed on my letter to the company's chief executive, who explained that the businesses in which they invest keep a record of 'accidents' but that only the deaths are reported back to London. He claimed that police investigations had been undertaken 'in almost all accidents involving fatalities', that compensation is treated 'on a case by case basis', but that 'information concerning compensation is confidential'. As regards prevention, he wrote that 'measures put in place to prevent deaths are case specific'. But I had clearly asked the wrong questions. The company later wrote to me explaining that they 'were saddened by my approach to this important issue' and believed that I was 'not committed to

74

working in a constructive manner'. That year its chief executive earned £380,000, of which more than £200,000 was paid in bonuses (Barnett, 2005).

The real price of roads

If roads are the cure for African poverty, we have learned nothing from history (Rodney, 1981). For centuries, Africa's roads have led to its impoverishment. In his economic history of Africa, Walter Rodney describes the role of Africa's transport infrastructure in this way:

> Means of communication were not constructed in the colonial period so that Africans could visit their friends. Nor were they laid down to facilitate internal trade in African commodities. There were no roads connecting different colonies or different parts of the same colony to meet Africa's needs and development. All roads and railways led down to the sea. They were built to extract gold or cotton and to make business possible for the trading companies and for white settlers. (Rodney, 1981)

So who does benefit from reducing the transportation cost of transnational trade? Global business revolves around resources, factories and markets. Raw materials are transported to factories where workers produce manufactured goods. These goods are then transported to markets so that consumers can buy them. If consumers are willing to pay more for the finished goods than it cost to produce them, the business will make a profit. And making a profit is what business is about. Cheap transport is good for profits because it reduces the costs of production and the costs of getting goods to markets. It also enables companies to take advantage of the lower wages of workers in poor countries. Indeed, cheap labour is one of the main reasons why the captains of industry are so excited about transnational trade. It is more profitable to set up factories in poor countries where wages are low than in rich countries where workers enjoy decent wages and standards of living. But poor people cannot afford to buy expensive manufactured goods and so the goods have to be transported back to markets in high-income countries. All this depends on cheap transport, which is bad news for road safety, physical activity and climate change, but good for profits (Roberts, 2004).

The globalization of trade leads to more freight, longer journeys, more road danger and more greenhouse gas emissions. In the poor countries that bear the brunt of the road death epidemic, trucks are responsible for the majority of crashes. In India, trucks are involved in half of crashes in cities and two-thirds of crashes on highways. The victims are mostly pedestrians and cyclists. Their experiences are part of the real social cost of international trade. And will small farmers in Africa compete with the subsidized grain from US agribusiness? Transnational trade is a new name for an age-old activity. For centuries, countries with greater economic and military power sought access to the resources and markets of weaker countries and millions died in the process. Rich countries claim that 'trade' benefits both rich and poor, but the historical record suggests otherwise. Walter Rodney believed that Africa's roads were built so that white settlers could make themselves rich at the black continent's expense (Rodney, 1981). The Uruguayan journalist Eduardo Galeano came to the same conclusion (Galeano, 1973). In his book *The Open Veins of Latin America*, Galeano wrote how the continent's transportation infrastructure was developed to drain its wealth into the ports and then out to the colonial economy. Nevertheless, publicly funded road building is only one of the three essential elements of a profit-centred transportation system. Cheap transport runs on cheap oil, and keeping oil prices within profitable limits entails huge public subsidies and masses of misery for a great many people.

The real price of oil

On the morning of 6 June 2007, Dr A's secretary handed him a letter. It was from one of his medical students. It contained a bullet and a threat, 'change the exam question paper and make it simpler otherwise you will be killed'. It was not his first death threat but this one took his breath away: 'Imagine those students will be doctors in the future and how much ethics they carry.' The doctor in question, who must remain anonymous for his own safety, is a physician in Iraq. He knows more than most about the real cost of oil. He has treated the mangled victims of US air strikes, children cut to shreds in marketplace bombings, as well as the daily deluge of routine severe violence.

In the week before the Anglo-American invasion of Iraq, 2 million people marched through the streets of London holding placards and chanting 'No war for oil'. In other capitals around the world, millions did the same. We now know that the US and UK governments lied about weapons of mass destruction. As a result of the war, hundreds of thousands of people have died, Iraq's infrastructure is in tatters and a culture of bloody violence reigns. The real cost of a litre of petrol is so much higher than the price that we pay at the pumps.

Going to war for oil is also expensive in monetary terms. According to former World Bank chief economist Joseph Stiglitz, US Congress will have appropriated more than $845 billion of public funds for its military operations in Iraq and Afghanistan

(Stiglitz and Bilmes, 2008). But this is a price that has to be paid because without its daily fix of 19 million barrels the US economy would experience painful withdrawal symptoms. The cost to the taxpayer of Britain's contribution to the military escapade is estimated at £20 billion. Heroin addicts use 'all necessary means' to feed their habit and often turn to crime. The shameful invasion of Iraq was nothing more than armed robbery, and according to leading international lawyers was every bit as criminal.

Americans comprise less than 5 per cent of the world's population but consume one-quarter of the world's oil. Most of the oil is used for road transportation. The USA woke up to the fact that it was well and truly hooked on oil in 1973 following the Arab oil embargo when a sudden hike in oil prices hit hard at the petrol pumps and for the first time US motorists had to consider carefully whether or not to use their cars. The need to safeguard US energy supplies led to the establishment of the US Department of Energy, and thousands of economists were set the task of ensuring that US oil demand could always be satisfied.

Economists have a lot of faith in market forces. They believe that when markets work perfectly supply and demand will balance beautifully as if an invisible hand were settling the scales. But the energy market is far from perfect. Its most important failure is that those who consume energy do not pay the full price. The real price would include the social costs of road death and injury, the human and economic costs of oil wars, the costs of obesity and diabetes, air pollution and community severance. It would include the cost to present and future generations of climate change. Indeed, Nick Stern, the economist who led the Stern Review on the economics of climate change, described global warming as the biggest market failure the world has ever seen. And of course if someone else is paying every time we fill up the petrol tank, the result is that we consume much more energy than we would if we had to pay the real cost. No rocket science here. If someone else is picking up the tab we feel much happier about overconsuming.

Economists at the US Department of Energy know all about the hidden costs of car use. They include the costs of environmental damage, the health risks and the so-called 'security costs of petroleum importation'. The 'security costs' are the military costs associated with 'defending' petroleum supplies in the Middle East region and have been estimated to range between $100 and $200 billion annually.

Cigarette packets carry health warnings – smoking can cause slow and painful death. The warning is written in large print so that smokers know the health consequences before purchasing. And if they do light up, two-thirds of the price of each cigarette is tax. The tax helps pay for the social costs of smoking. Oil addiction also causes painful death, and one way to tackle it would be to make sure that motorists pay the full cost every time they fill up and that they know exactly what they are paying for. For the markets that Western politicians esteem so highly to work their magic, every dead Iraqi child, every exploded soldier and journalist, every murdered checkpoint victim and bombed-out marketplace would have to be priced and paid for at the petrol pumps.

Cars

Once an extensive, publicly funded road network is in place, studded with tens of thousands of petrol stations selling fossil fuels at bargain prices, with millions of rent-free parking places expropriated from the civic space, the scene is set for the ascendancy and dominance of the motor vehicle. When the first Model T Ford rolled off the assembly line in 1908 it was a miracle of mass production. In the first decade of that century, car registrations in the USA skyrocketed from 8,000 to almost 500,000. Getting Americans hooked on cars was the name of the game. In 1932, General Motors bought up America's tram system in order to close it down. To sell cars they had to create distance. The private passenger car was marketed as an escape route from dirty, overcrowded cities to leafy garden suburbs, and the town planners colluded. US cities like Los Angeles, Dallas, Houston and Phoenix were moulded by the passenger car into vast urban sprawls that are so widely spread out that it is now almost impossible to service them economically with public transport (Roberts, 2003). As the cities sprawled outwards, the car industry consolidated. Car-making is now the main industrial employer in the world, dominated by a fist of five major groups, of which Toyota and General Motors are the largest.

The automobile industry is now the keystone of the US economy. What's good for General Motors is good for America and what's good for General Motors is selling cars. But the car industry is struggling, suffering a serious crisis of overproduction and surviving on the smallest of profit margins. It is crucial that every branch of government, from the Treasury to the Department of Defense, does its bit to support car sales both at home and abroad. If the car industry needs a handout from the US taxpayer, then that is exactly what it gets. It is for this reason that American presidents board planes to make friends with human-rights-abusing despots in distant countries whilst denying the existence of man-made climate change. It is for this reason that World Bank economists are sent out from Washington like missionaries, going to Africa, Asia and Latin America, spreading the gospel of road building.

The politics of roads, oil and cars also explains, of course, why the USA has one of the fattest populations in the world, with a carbon footprint of elephantine proportions.

The car makers understand the threat posed by the great mass of the walking public. At the moment, they are busy manoeuvring themselves into pole position in road safety politics. They have to ensure that efforts to improve global road safety don't adversely impact on car sales. In 2006, the Fédération Internationale de l'Automobile set up a Commission for Global Road Safety with a remit to 'examine the framework for and level of international cooperation on global road safety and to make policy recommendations'. At its inception, the Commission was chaired by former UK defence secretary Lord Robertson, and had eight commissioners, one from each of the G8 group of wealthy nations. If you wanted to represent the interest of motoring classes you could not put together a more able group of commissioners. Canada was represented by an executive director at General Motors, Japan by a board member of the Bridgestone Corporation, the major transnational tyre maker. Russia was represented by the president of the Russian Automobile Federation and Italy by a former president of the Automobile Club of Italy. Michael Schumacher represented Germany, and France was represented by Gerard Saillant, a doctor who works on the medical aspects of Formula One. The UK commissioner was the chief economist at Lehman Brothers, the US investment bank whose later collapse precipitated the perfect storm of global economic chaos. The Commission's patron is Prince Michael of Kent, a former racing driver, now a member of the British Racing Drivers Club and the Bentley Drivers Club. Lord Robertson himself was then deputy chairman of the Board of TNK–BP, a Russian oil company. According to the House of Lords' Register of Interests, which shows that the FIA paid Robertson to attend the 2006 Monaco Grand Prix, the Commission holds its meetings at the race track (Roberts, 2007).

Working through the Commission, the FIA and the car lobby are taking a lead role in global road safety. They would like to set the policy agenda for road safety and thus gain considerable influence in global transport policy. They do not want to fund road safety efforts but to dictate how other organizations spend their money, and in particular how development money is spent. Former World Bank president Paul Wolfowitz was eager to confirm the Bank's willingness to implement the Commission's recommendations, and former UK secretary for state for international development Hilary Benn welcomed the Commission's proposals.

Although most people in poor countries will never own a car and most of the victims of road traffic crashes are pedestrians, the Commission has worked hard to ensure that the views of the motoring elites dominate transport policy decisions. Unelected, with only token representation from developing nations, the car lobby wants to dictate how poor countries' governments spend the development loans that their impoverished people will repay for decades. The Commission strongly endorses road building in poor countries, claiming that this is essential for achievement of the Millennium Development Goals. The available evidence suggests otherwise. The buried DFID review on transport and development concluded that high levels of private car use were not 'financially, socially, or environmentally sustainable in a developing country context'.

The car lobby's favourite road safety policy is pedestrian education. Despite decades of evaluation research, safety education has never been shown to reduce road injury rates, a point emphasized by the World Health Organization (WHO) in the

World Report on Road Traffic Injury Prevention (Peden et al., 2004). Road user education is favoured by the car lobby because it places the responsibility for road traffic injury squarely on the victim and has no impact on industry profits. Its primary purpose is ideological. It sends the message that the road space belongs to drivers and that pedestrians and cyclists must look out or die. This also applies to children, by the way, who account for 300,000 of the 1.2 million road deaths each year. Awareness campaigns are another favourite (Duperrex et al., 2002). The Commission promoted the 'Think Before You Drive' campaign, supported by the Bridgestone Corporation, which reminded drivers to use child seats and seatbelts and to check their tyres. Sensible suggestions they may be, but such exhortations have no discernible effect on road safety. On the other hand, the campaign may improve Bridgestone's corporate image.

At first sight there appear to be many different stakeholders in the global road safety policy arena, but careful examination reveals otherwise. In 1999, the World Bank – arguing that a partnership between businesses, NGOs and governments can deliver road safety improvements in poor countries – established the Global Road Safety Partnership (GRSP), a business partnership that includes the automotive giants General Motors, Ford, Daimler Chrysler, Volvo and drinks multinationals such as United Distillers and Bacardi–Martini. General Motors was represented on the GRSP and on the Commission for Global Road Safety.

A 2006 study compared the frequency of use of different road-safety-related words in GRSP road safety reports and in the *World Report on Road Traffic Injury Prevention*, a report that was prepared relatively independently of business concerns by the WHO. In the GRSP reports there was a clear lack of reference to pedestrians and cyclists. In the WHO report, 'speed limit' occurred seventeen times in every 10,000 words; in the GRSP reports, just once. 'Pedestrian' was used sixty-nine times by the WHO, and fifteen times by the Partnership; 'buses' and 'cyclists' were mentioned thirteen and thirty-two times, respectively, by the WHO but not once by the Partnership (Roberts et al., 2006).

To avoid catastrophic climate change and prevent the seemingly inexorable upward shift in the average BMI of the human race, we must reduce the amount of dangerous kinetic energy in our towns and cities. Human beings must use human bodies for human movement. We must reclaim our streets for walking and cycling, for games and gardens, for children's play and for street parties, and for whatever other public purpose our creative imagination can find for them. Real road safety means reducing road danger, which implies far fewer motor vehicles travelling at much lower speeds. Were the automobile industry to endorse such a vision for the future it would be tantamount to voluntary corporate euthanasia. Of course, the car industry can and does champion seat belts (although they resisted doing so for many years), helmets (even for pedestrians) and road safety education (despite the scientific evidence that it is ineffective) but they will never call for car-free cities based around walking, cycling and public transportation. Nor, for that matter, will any other transnational businesses whose profits depend on underpriced road transportation, as I discovered after a meeting on climate change in Madrid in November 2008.

Some hard facts about a soft drink

'What are you doing now? There is someone I want you to meet.' I have just walked out of the morning session of a World Health Organization consultation to define a research agenda climate change and health and two WHO staff members are ushering me out of the building. During the three-minute walk between the Spanish Ministry of Health and the upmarket restaurant where we are to have lunch, they explain that we are about to meet representatives from Coca-Cola who want to fund research on physical activity. The Coca-Cola people are waiting for us. Around the table are the chief scientific and regulatory officer, the director of corporate social responsibility, and the director of institutional relations and communications from Coca-Cola Spain. For the next thirty minutes we discuss the links between physical activity and health. It seems that Coca-Cola want to fund research on physical activity, but the WHO does not feel that it can accept money from the beverage drinks industry and so they want my university to take the money instead. Their chief scientific officer is flying back to the USA in a few hours' time and has to leave for the airport. There are smiles and handshakes and the meeting ends.

A few hours later, there is an email message waiting for me, 'Great to meet you in Madrid. Brief meeting but very promising … we will be delighted to work with LSHTM in this fascinating subject of climate change prevention and health improvement through physical activity.' The next day a research proposal lands in my inbox with a message saying how pleased they are to work with my university. I could only presume that I was expected to approve the proposal and send it back to the company and wait for the funding.

The Coca-Cola people said that they wanted to get people active. It is not surprising that a company selling an energy-dense soft drink would want to focus attention on the output side of the energy balance equation. It is clearly more convenient to Coca-Cola to lay the blame for global fatness on sloth, rather than on the 35 grams of sugar inside a 330 ml Coke can. Burning off the 139 calories it contains would require a generous serving of physical activity and for many people leaving the can on the shelf would be the easiest way to avoid gaining the extra weight.

Coca-Cola might have thought that promoting physical activity was a safe area for company-sponsored 'research'. However, this is the case only if they ignore the health and environmental effects of a global distribution network that delivers a sugary drink to almost every corner of the world. The secret of Coca-Cola's commercial success is not its fiercely guarded secret formula but its ability to externalize the real costs of transportation. So I offered Coca-Cola a research proposal of my own. I suggested a research project about the road deaths and injuries associated with Coca-Cola's road distribution network, the effect that the resulting road danger has on physical activity, and the climate-change impacts of the fossil-fuel energy use its road distribution network depends on.

Coca-Cola cannot claim that road deaths and injuries are not a problem for the company. According to the 2006 Social Responsibility Report of its European bottling operation, there were nineteen fatal work-related accidents in 2006, most of which were road deaths, and most of which were in poor countries. The report did not give the number of non-fatal injuries. Nor did it mention the number of injuries to people who do not work for the company. This is important for road freight in poor countries, where pedestrians and cyclists, many of them children, account for a large share of road crash victims. The report did mention that a defensive driving training scheme had been launched in response to the high number of road deaths among company workers in Nigeria. This may be good PR but research shows that these programmes are useless in preventing crashes (Ker et al., 2005). After I sent my alternative research proposal to Coca-Cola, all communications with the company dried up (Roberts, 2008).

Happy-hour oil

This book is about fossil fuels and how the energy they contain has shaped our bodies and our environment in a way that now presents a threat to our health and sustainability. Working back along the chain of causation, the causes have become increasingly intangible and it might seem preposterous to claim that the injured children who were etched onto my memory as a trauma doctor have any connection with how corporations do business and the price of oil. But this is exactly what this book claims. It is also what the road death statistics show. In 2008, the US secretary of transportation congratulated herself on the record low number of US road deaths in that particular year. Only 31,110 people had died on her nation's roads from January through October, compared to 34,502 over the same period in 2007. She claimed the decline was due to her department's focus on safety. The real reason was the record high oil prices in 2008, which cut the amount of kinetic energy on US roads. Rather than celebrating the decline, world leaders in response to the clamour from business were hopping on planes bound for Saudi Arabia in order to persuade Saudi princes to increase oil production in order to get the prices back down again.

If business had to pay the true costs of transport, international trade would be inefficient and there would be much less enthusiasm for it. Fortunately for business, the public and the global environment pick up the bill so that transnational trade can remain lucrative. And by ignoring the social and environmental costs of transport, the world has vastly overconsumed fossil fuels. Our planet has enjoyed a century of happy-hour oil, but its atmosphere is now seriously intoxicated and its patrons are dangerously overweight.

5

Contract and converge

The first patient that I actually observed being enrolled into the CRASH 2 clinical trial was a Colombian soldier who had stepped on a landmine in the mountains north of Medellín. Although by that time over 12,000 patients had been enrolled, this was the first time that I had witnessed the process. Two paramedics wheeled him into the emergency room and then turned and left, as though in a hurry to get out. Moments later his mother, having convinced the security guards that this was her son, tiptoed into the room to stand by his side. Whether she had been with him on his four-hour ambulance journey, or else had been waiting for him at the hospital, I do not know. The soldier was lean and cinnamon brown, probably still in his teens, with his hair cropped so closely around the side of his head that it left just a small black runway on the top. He looked more puzzled than in pain as the nurses cut off his trousers to allow the trauma surgeon to examine the ragged stub of his left leg, but when her gloved fingers started probing the meat of his thigh he stiffened and arched. Within minutes his mother had given consent for him to be entered into the trial, the entry form had been filled in, and the trial treatment had been started.

Worldwide, most patients with severe internal bleeding have been on the receiving end of motor vehicle kinetic energy, but in Colombia bullets and landmines account for a large proportion of the patients. Clinical trials in trauma care have to be simple because they are carried out in emergency situations. We hope that the new treatment will save lives by reducing blood loss but the only way to find out is to recruit thousands of bleeding trauma patients, give one half the new treatment, give the other half a placebo, and then later compare the death rates in the two groups. To build a collaboration of trauma doctors from all around the world, the trial aims and methods must be simple and clear.

Climate change is a slow-burn global health emergency and there is an urgent need for an international agreement on how greenhouse gases should be reduced. The foundation for any such an agreement will have to be simple, science-based and transparent. There are of course complicated scientific and political issues at stake, but time is running out, and unless the arguments can be made clear and transparent what can be the prospects for success? Complexity is a strategy used by professional elites to maintain control. Proclaiming that a problem is complex is shorthand for saying that that you have no role in solving it. For example, referring to the need for an international agreement to prevent climate change Tony Blair wrote:

> Given the complexity of the issues involved, the imprecision of much of the data, and the extraordinarily tricky interplay between the political, the technical and the organizational, answering the question of 'how?' is as difficult as any the

international community has grappled with since the design of the post-war Bretton Woods economic institutions. (Blair, 2008)

Does this sound like someone who wants the public to have any say in how we reach a binding climate-change agreement? But there has to be public support for such an agreement if we are going to avoid dangerous climate destabilization.

The earth is getting hotter. Reducing our personal carbon footprint might make us feel virtuous and we will probably shed some excess weight, but we should not delude ourselves that it will save us from catastrophic climate change. What we do as individuals will never be enough. Mayer Hillman, author of *How We Can Save the Planet*, makes this point with humour, joking that he is saving up to buy a house in Mayfair (Hillman, 2004). To raise the necessary funds, every evening when he comes home, he puts his loose change in a jar by the door. Yes the jar gets a little bit fuller by the day, but he will never save enough money to buy his dream house. Moving in the right direction does not necessarily mean that we will arrive at our destination. And of course what we do is not the whole story. Even if we live the most carbon-frugal lives possible, unless the rest of the world changes too, we will fail to prevent the steady accumulation of greenhouse gases that is endangering the planet.

We have seen that the causes of population weight gain are environmental not personal. Even if we do manage to walk and cycle everywhere we go, and purge our homes of energy-dense foods, unless everyone does the same, we will always be swimming against the current. And the direction of flow is clear. The population is getting fatter. Short term, we might summon the will to make the necessary sacrifices. By moving more and eating less we can rapidly lose weight, but the body responds to sudden weight loss by increasing appetite and reducing energy expenditure, and any weight lost is likely to be quickly regained.

This chapter outlines a simple and transparent policy framework for reducing greenhouse gas emissions. If such a framework was implemented, not only would it ensure that atmospheric levels of greenhouse gases were kept to within a safe level but all of the steps suggested in the earlier chapters of this book would become the easy options, as they would for everyone else around us (Egger, 2008, 2007). The policy is called Contraction and Convergence, a global climate policy framework proposed to the UN in 1990 by the Global Commons Institute (GCI, 1990).

Contraction and Convergence

Contraction and Convergence is a simple science-based starting point for an international agreement on reducing greenhouse gas emissions that is based on the principle of justice and equity. The amount of greenhouse gases in the atmosphere is increasing rapidly. The concentration of carbon dioxide, the chief greenhouse gas, is now about 387 parts per million (ppm), which is nearly 40 per cent higher than the concentration at the start of the Industrial Revolution (280 ppm), and higher than at any time in the past 650,000 years. To ensure our survival, the concentration of carbon dioxide must be kept within a safe upper limit. Because carbon dioxide accumulates in the atmosphere, where it remains for about a century, in order that the atmospheric concentration does not rise any higher the amount of carbon dioxide that is emitted must fall, eventually reaching very low levels, such that emissions just balance natural loss. This is the contraction part. Establishing a safe upper limit is a technical matter for climate scientists; although there is bound to be heated debate, it should be possible to reach agreement on where the limit is set and a date by which this concentration should be reached. This upper limit will determine the amount of greenhouse gases that can be emitted in the future. The key question that remains is who should be allowed to emit them.

From space, the life-sustaining gases of the earth's atmosphere are a thin blue layer enveloping the planet. The idea that anyone should own the sky, the silent space where the clouds soar, is absurd. The atmosphere is a global good that all the citizens of the earth must share. It follows that when it comes to parcelling out entitlements to emit greenhouse gases, everyone should have an equal share. Currently, the emission of greenhouse gases is far from fair. The average per capita emission of carbon dioxide in the USA (20.6 tonnes) is about sixteen times higher than that in India (1.25 tonnes). Contraction and Convergence sets out a timetable for when per capita emissions should converge to equal per capita shares. This is the convergence part. The policy is a compromise. It acknowledges that everyone has an equal right to the atmosphere but recognizes that the wealthy world will need some time to make the transition to fair shares. In the convergence phase, wealthy countries will have to make cuts even as emissions from poorer countries are increasing. However, once per capita emissions converge, rich and poor alike will have to reduce their emissions together. The policy also allows for emissions entitlements to be traded, which should help to ease the transition to equal shares whilst ensuring that the safe upper limit is not exceeded. Wealthy countries with high carbon dioxide emissions will have to buy the unused entitlements of poor countries, resulting in a transfer of wealth from the rich world to the poor.

In high-income countries, emissions by individuals account for about 40 per cent of total carbon emissions, with the rest coming from factories and businesses, which means any corporate cap-and-trade system will need to be accompanied by personal carbon rationing (Hillman and Fawcett, 2004). With personal carbon rationing, every year, each individual would be allocated an equal number of tradable carbon rations. The size of the national carbon budget would be set by an independent organization that links national obligations to cut greenhouse gas emissions to global emissions policies (Hillman and Fawcett, 2004). The extent of the personal carbon ration would be determined by the national budget. The important part is that setting a national and personal carbon ration based on what the planet can stand guarantees that we live within our environmental limits. Of course, if the population of a country grows then everyone's personal carbon ration would get slightly smaller. Population and sustainability are inextricably linked. If there are more people, the finite share of the right to put carbon dioxide into the atmosphere has to be shared between more people. It could not work any other way without leading to ecological destruction.

Under such a system of carbon allowances, people would hand over their carbon credits whenever they buy fossil-fuel energy, such as fuel for transport or heating. One can imagine that the technology already in place for direct debit and credit cards could be adapted for use in personal carbon trading. People who lead highly carbon-intensive lifestyles might use up their allocation of carbon credits and would have to buy the unused credits of those living more carbon-frugal lives. Because greenhouse gases accumulate in the atmosphere, where they remain for years, to stabilize concentrations within a safe limit the allocation of carbon credits would have to be reduced year on year, eventually down to zero. As a result, the cost of buying unused credits would increase as they become increasingly scarce. There would be a strong personal incentive not to waste carbon allowances and a strong political incentive to reduce the carbon intensity of society.

Under a system of carbon trading there are likely to be winners and losers. Globally and nationally the carbon-profligate would lose out and the carbon-frugal would gain. A 2008 study by the UK Department for Environment, Food and Rural Affairs (DEFRA) found that because low-income households tend to have lower carbon emissions than high-income households, the poorest sections of society would win, whilst the more wealthy sections would lose out (DEFRA, 2008). When homes are ranked into tenths according to their household income, the DEFRA study found that 71 per cent of households in the lowest three income tenths would have leftover allowances that they could sell, whilst about half of households in the highest three income tenths would have to buy unused allowances or cut their emissions. Transportation research by the University of Oxford shows that carbon emissions related to travel are steeply socially stratified, with the rich being massively more polluting than the poor.

Under a system of carbon rationing, there would be a strong incentive to reclaim the streets from motor vehicles. It is unlikely that people would want to waste valuable carbon credits on foolishly short car trips that they could easily walk or cycle and there would be a strong demand for safer streets. Public demand for improved road safety might lead to much greater investment in infrastructure for safe walking and cycling. Legislation could be passed to give pedestrians legal right of way, as is the case in some streets in the Netherlands, and that Home Zones, which are residential streets with very low vehicle speeds and pedestrian priority, might become the rule rather than the exception. People might regain their sense of distance, and their physical fitness would start to improve. They would realize that most of their access needs can easily be met by bicycle or by public transport. Market forces would eventually ensure that those access needs that are not met within a cycle ride soon would be. As population fitness levels increase, half an hour of cycling morning and evening might be just as acceptable to city commuters as is the largely wasted time spent commuting by car or train now. Purposeful movement would become the key to maintaining a healthy body weight and the fitness industry would go into decline.

Sedentary home-based activities such as watching television, playing computer games or surfing the net might become less popular as people become reluctant to use up carbon credits on home heating and electricity use. People would begin by insulating their homes and wearing warmer clothes. The street space would become a public space, and public activities would replace sedentary solitary indoor pastimes. At the same time corporate carbon trading will mean that fossil-fuel-intensive, energy-dense, processed foods would become more expensive than locally grown seasonal fruit and vegetables. Healthy eating will become the easy option.

We cannot know, nor do we need to know, the extent and the diversity of the strategies that individuals and society might develop to reduce fossil-fuel energy use. But we can predict that when the creative intelligence of the world is directed towards this aim, the future will be a different place. Our overconsumption of fossil-fuel energy has given us climate change and pandemic fatness. Contraction and Convergence by weaning us off fossil fuels will protect the planet and release its inhabitants from the chains of corpulence (Egger, 2009).

The principle of Contraction and Convergence has been endorsed by governments, non-governmental organizations, environmentalists, scientists and religious leaders around the world. You can get some idea of the range of individuals and organizations that support the principle from the Global Commons Institute website. It reads like an international Who's Who of the great and the good. In fact, according to the UK government, the only group that is not convinced about the merits of the approach is the public. In 2008, the UK government published the results of its 'pre-feasibility study into personal carbon trading'. It's odd given the support for Contraction and Convergence that the government should conduct a 'pre-feasibility study' implying that it was not even sure that it is worthwhile proceeding to a feasibility study. Nevertheless, the study concluded that personal carbon trading 'is an idea currently ahead of its time in terms of its public acceptability and the technology to bring down the costs'. In other words, they are claiming that we are not ready for it and that it is too expensive.

You might be surprised to find that in the view of the government you are the main obstacle to moving ahead on this critically important issue. I suspect that many people will never have heard about the proposals. This is not your fault. Given its importance, the media coverage of carbon rationing has been scant at best and the amount of airtime given to it by our political leaders has been rock-bottom low. They blame you for being killed on the roads, they blame you for getting fat, and they will blame you when the planet fries. The costs of setting up the scheme for the UK were estimated at between £700 million and £2 billion. These are large sums of money, but must be considered in the context of the tens of billions of pounds of public funds that were spent to prop up failing financial institutions. Saving the planet must surely be allocated as much importance as saving banks.

Governments were quick to bail out the banks because the people who run the banks and our political leaders are friends.

When wealthy people start declaring a crisis and the need for an urgent state-funded solution, the media make sure that we get to hear about it and our elected leaders respond. Not so when it comes to the environment. This time wealthy people have more to lose. Burning fossil fuels irrespective of the consequences makes a lot of money for some. At the individual level, the people drinking champagne in the business-class lounges of international airports are not your average punter; they are the wealthy elite. The media are gagged due to their financial dependence on advertising income from motor vehicle manufacturers, air travel and distant holidays and are understandably reluctant to bite the hand that feeds them. So it remains a secret, and the main excuse for not taking it forward is that the public are not ready for it.

Government denial of the inevitability of carbon rationing means that when such a system is finally introduced, as it must be, the emissions cuts that will have to be made will be that much steeper. Because carbon emissions eventually have to be reduced down to zero, the total carbon budget for any particular country will have to be reduced year on year. Shunting the decision about carbon rationing into the future will mean that far more radical emissions cuts will have to be made and that the thinking time for finding creative societal solutions will be that much shorter. It will also mean that the population will be that much fatter, heavier and slower, making the modal shift to walking and cycling even more difficult.

One unfortunate feature of the policy of Contraction and Convergence is its name, which implies both austerity (contraction) and uniformity (convergence). We know that greenhouse gas emissions must contract if we are going to prevent ecological disaster. But greenhouse emissions are not the metric with which we judge the quality of our lives. It is not what we consume or how much we pollute but how happily we live our lives that matters the most. Being physically active, connecting with the people around us, contributing to the wider community and having trees, bushes, flowers and birds around us are the foundations of well-being, and all of these will expand in a low-carbon society. The freedom to move, in a safe and quiet environment, breathing clean air, without the fear of road danger or crime, is not austerity.

We also know that greenhouse gas emissions have to converge to equal per capita shares. The first article of the Universal Declaration of Human Rights is that 'all human beings are born free and equal in dignity and rights'. No individual or society can legitimately claim a greater right to pollute the atmosphere. However, equality does not constrain diversity. Quite the opposite, it allows for its fullest expression. Currently, because it serves the interests of the businesses that make up the petro-nutritional complex, all roads lead to the shopping mall and all the malls look the same. They contain the same retail chains and sell the same mass-produced products. And to channel the wealthy people to the malls, supermarkets and petrol stations more efficiently, everyone else is forced out of the way. Currently, throughout Asia and Africa, a miscellaneous multitude of small traders is being cleared off the streets to make way for road expansion, flyovers and shopping centres, the hardware of what is currently considered to be economic development. Professor Dinesh Mohan from the Indian Institute of Technology believes that this is bad for safety and bad for the economy:

> Hawkers have to be there on roads. They are the most enterprising entrepreneurs who are doing business every day by occupying little space, consuming very little energy and generating reasonable revenue. They are contributing to the

economic growth. It is because of them that the Indian roads are comparatively safer from mischief doers. (Stealth of Nations, 2009)

Roads and cars channel people and profit past the small businesses towards the big business, from multiplicity towards monopoly. Carbon rationing would foster diversity between, as well as within, nations. Under Contraction and Convergence and carbon trading it would not be in the interests of carbon-thrifty nations to follow the same fossil-fuel-intensive paths to 'development' that were taken in high-income countries. If Africa and Asia became carbon copies of North America, they would have killed the golden goose. Low-carbon economic and human development will not look like anything that has gone before. Real human development is about the expansion of human freedoms, and freedom brings diversity not austerity. We will take a closer look at what a low-carbon world might entail in the next chapter.

6

The era of the bicycle

Cuba, 1990. From a barnacled outcrop of sea-lashed limestone that points like an arthritic finger into the blue-grey waters of Havana bay, fish-slim Cuban children fling themselves into the waves. In Santa Clara, a sugar-cane cutter brushes a calloused finger over the blade of his machete while he waits for the rusty tractor that is lumbering like a steel skeleton through the cane field towards him. The Cuban Revolution has weathered three long decades of struggle. But on the other side of the world a geopolitical storm is brewing; when it eventually makes landfall in Cuba, it will leave the children hungry and the tractors silent.

If US policy on Cuba could be likened to schooling, with the headmaster at his desk in the Oval Office, then Cuba is the badly behaved pupil. Since the early days of the Cuban Revolution, the slender island state had been punished for its disobedience of the USA by the imposition of a strict economic embargo. This it had managed to endure, but only because of its extensive trade links with the USSR. Trading Cuban sugar for Soviet petroleum had kept the island's economy alive, but in 1990 the Soviet superpower was teetering on the brink of collapse, and when it finally fell apart Cuba was thrown into a brutal economic crisis.

In the years following the revolution, 1 tonne of Cuban sugar bought 8 tonnes of petroleum. After the fall of the Soviet Union, it bought less than 2 tonnes. Low sugar prices on world markets and a lack of foreign exchange meant that Cuba could no longer import enough petroleum to keep the wheels of its economy turning. The USA wasted no time in exploiting Cuba's vulnerability. At a 1993 summit between the Russian president Boris Yeltsin and US president Bill Clinton, a key condition on the finance package offered to support the crumbling Russian state was the cessation of all petroleum exports to Cuba. Cuba's 'Special Period' had begun.

In the coming years, Cubans would suffer severe shortages of food and fuel. Vitamin deficiencies would cause over 50,000 cases of a rare neurological disorder. The slow upward drift in the average Cuban BMI would be abruptly reversed as shortages of fuel and fertilizer for use in farming constrained food energy production, and transportation fuel shortages increased energy demands. Cuba was down but not out. As transportation ground to a standstill, Castro announced its salvation. Referring to the Cuban 'Special Period' Castro explained 'it also has its positive sides and one of those is that we are entering the era of the bicycle.' The following year Cuba bought a million bicycles from China and sold them to students for 60 pesos and to workers for 120 pesos. It built the bicycle factories that would assemble about half a million cycles over the next five years. With the US embargo throttling the economy, and its lifeblood of Russian petroleum ebbing away, bicycles were the saviour of the Cuban Revolution.

Castro's solution to the Cuban energy crisis had a sound scientific basis. The bicycle is the most energy-efficient mode of land transportation that exists. Cycling burns about 35 kilocalories of food energy per mile, whereas walking the same distance burns three times as much. By comparison, car travel uses about 1,860 calories of fossil fuel energy per mile. Without bicycles, Cuba's nutritional and transportation problems would have cut much more savagely than they did.

Bicycles and development

Despite its relative poverty, Cuba has an international aid policy that would put many Western governments to shame. Cuba trains thousands of doctors and sends them to the parts of the world where they are needed most, but especially to Africa. For most Western governments, on the other hand, African development aid means building highways between the mines and the ports. The lessons learned during the Cuban era of the bicycle would have important implications for transportation in Africa. Across large parts of sub-Saharan Africa, the most common form of transport is 'the legs, heads and backs of African women and children' (Peters, 2000). The lion's share of transport in Africa is in rural areas, on unpaved paths and tracks, carrying crops from the fields to the villages. About 70 per cent of the population targeted by the Millennium Development Goals (MDG) anti-poverty initiative live in rural areas. Unfortunately, most travel surveys count only cars and trucks and ignore the daily drudgery of walking from the village to the fields and back again, carrying head-loads of crops, typically weighing 30 kg or more over an average distance of about 5 kilometres. Between 70 and 90 per cent of the work of carrying crops is done by women and children, who also do almost all of the carrying of water and firewood (Peters, 2001). The loads are carried either on the head or on the back (Riverson and Carapetis, 1991). A study in Addis Ababa of women fuel carriers found that the average load weighed 36 kg (75 per cent of body weight), and was carried for around 12 km (Haile, 1989). Close to 20 per cent of women carried loads heavier than their body weight. Not surprisingly, back and joint pains are common and there are high rates of miscarriage.

Bicycles with or without trailers could dramatically reduce the amount of time and energy women and children spend walking and head-loading. Spending two hours per day walking to and from the fields seriously reduces the time available for agricultural, domestic and social activities. Greater use of bicycles could improve access to education and health facilities. The infrastructure needed would be much less expensive than for motorized transport. A 2 metre-wide unpaved bike trail would cost less than 10 per cent of the cost of a 6 metre-wide rural road for motor vehicle use. I have already referred to the research conducted by University College London that concluded that there was no evidence that road building promotes economic development or poverty reduction. The report did, however, emphasize the importance of bicycle use (Banister et al., 2005). One study in the Makete District of Tanzania found that, whereas building a feeder road saved households 120 hours per year, investing in a bicycle saved the family 200 hours per year (Sieber, 1997). A study in Ghana found that a two-person day trip to move 1 tonne over a kilometre could be done with a bike trailer in under an hour, and a 'before and after study' of subsidized bicycle provision in Uganda found a range of beneficial effects, including more frequent trips to markets and health-care facilities and increased household income (Banister et al., 2005). It would not be prohibitively expensive to increase cycling in Africa. For the cost of a single highway in Senegal, every person in that country could be provided with a bicycle (Banister et al., 2005).

The distribution of bicycles to African health workers has been shown to cut journey times, reduce transport costs and increase the numbers of patients visited. The provision of bicycles to nurses providing care for terminally ill patients in South Africa enabled them to reach fifteen times more patients. Non-motorized transport is also effective in generating employment. Cycle rickshaws in India provide an estimated 6 to 9 million jobs, and in China more people are employed making bicycles than in making cars (Banister et al., 2005).

The week after I got back from the Bloomberg meeting in New York, the UK government announced an agreement with eight African countries and the World Bank that will see the construction of 5,280 miles of roads in Africa (McGreal, 2009). The British public will invest about US$100 million in the US$1 billion project that will focus on linking inland trade hubs with ports in Tanzania, Mozambique and South Africa. According to Gareth Thomas, UK minister for international development, 'Africa has a wealth of fantastic produce to offer UK consumers and the rest of the world, from tropical fruits and vegetables to minerals and gemstones but is held back from being able to compete and deliver by this poor trading route.' He added that the Department for International Development is working 'to clear a path for improved prospects for the lives for Africans and provide better choice for consumers' (DFID, 2009). The UK government's plan was to 'revolutionize' trade routes in Africa through a road-building programme that would link landlocked African countries to the ports. According to newspaper reports, Thomas told the conference that high transport costs were a major reason for African poverty and that 'without a better route out, supermarkets and traders are missing out on quality produce and African communities are missing the chance to trade, create more jobs and in turn feed themselves and school their children.' Writing with no hint of irony, Chris McGreal, the *Guardian*'s Africa correspondent, described the extent of the initiative as 'on a scale unseen since the road and rail networks were first laid down by European Colonisers'. His comparison is eerily appropriate bearing in mind that the roads and railways linking African mines to the ports were the 'open veins' through which the continent's wealth bled out into the colonial economy. Clearly, infrastructure for walking and cycling is not the sort of infrastructure that interests the development ministers of wealthy nations when they come to dispense their foreign 'aid'.

Both the development minister and the *Guardian*'s Africa correspondent appeared to have overlooked the adverse public-health impacts of African road-building schemes. According to the World Health Organization, road deaths in Africa could double between 2008 and 2030. Road traffic crashes already account for more deaths than from malaria, and by 2030 road death is expected to exceed AIDS as a cause of death world-wide (WHO, 2010b). One in five road deaths in Africa involves a child between five and fourteen years old. In some African countries, the growing burden from road traffic crashes has even started to impact on the delivery of other public health programmes. Doctors in Ghana were forced to stop a programme of prenatal home visits, an initiative that had been launched in an attempt to reduce maternal mortality, because road traffic victims were taking up so much of their time and health-care resources. According to Dr Dodi Abdallah, a doctor at Winneba Hospital, situated on the 15 km stretch of road between Accra and Winneba, known locally as Ghana's Bermuda Triangle because of the large number of people who lose their lives there, 'we are losing the battle against maternal mortality because of the sheer pressure of accident emergencies' (allAfrica, 2009). On 25 May 2009, Dr Tajudeen Abdul-Raheem, UN Millennium Campaign deputy director for Africa was killed in a car crash in Nairobi whilst on his way to launch a maternal health campaign in Rwanda. He died on African Liberation Day. Road traffic crashes are estimated to cost developing countries $53 billion per year, which is more than they receive in development aid (WHO, 2004).

On your bike

Cycle-based transport should benefit the citizens of rich and poor countries alike. Unlike the demand for food and shelter, there is no intrinsic demand for transport. Instead, transport is a 'derived demand', arising from our need for access. We demand access to employment, education, health care and recreation, and in cities everywhere cycling can provide this access in a healthy and sustainable manner. Because cycling is a far more energy-efficient method of transportation than walking and can make carrying heavy loads much less exhausting, it can help the undernourished in poor countries to gain weight. On the other hand, because cycling consumes more food energy than driving a car, it can help fat people in rich countries to lose weight. Cycling is liberating, fun and healthy, and by replacing motor vehicle use it will reduce greenhouse gas emissions.

For many people, the first reaction to the last paragraph will go something like 'I would cycle if only it was not so dangerous.' It is completely understandable that people should respond in this way. The road traffic injury risks to cyclists are all too apparent.

However, we are much less aware than we should be of the health risks of not cycling. We were designed to move and not to stand idle. Physical inactivity substantially increases our risk of heart disease, stroke and cancer. As a result of our sedentary lives, the slippery smooth walls of our blood vessels steadily accumulate a fatty sludge called atheroma, which can eventually lead to the blockage of blood vessels, halting the flow of blood, starving the tissue of oxygen. If this happens in the blood vessels feeding the heart wall, the result is a heart attack; a blockage in the blood vessels supplying the brain with oxygen causes a stroke; and both can lead to death or severe disability. Physical inactivity accelerates atheroma accumulation. Not moving is also dangerous.

The overall risk of death for adults who cycle to work on a regular basis is between 10 and 30 per cent lower than for those who drive to work (Woodcock et al., 2009). This survival benefit persists after controlling for a range of factors that might differ between cyclists and motorists. In other words, even taking into account road danger, the balance of health risks and benefits is strongly in favour of cycling. Cycling in traffic may look dangerous but not cycling is more dangerous. There are consistently fewer deaths than expected from heart attacks, strokes and cancer among cycle commuters. Take a fifty-year-old office worker who drives 4.5 km to work every day, a distance that is easily manageable by bicycle. If he switches from driving to cycling for 90 per cent of these trips, at the end of the first year he could have shed up to 5 kilos of adipose tissue and will enjoy a 20 to 40 per cent reduction in the risk of premature death and a 30 per cent reduction in the risk of type 2 diabetes. He will also have better mental health and improved fitness (Woodcock et al., 2007).

There is also evidence that the injury risks for cyclists decrease as more people take up cycling. Per kilometre, cycling is safer when there are lots of other cyclists around. It has been estimated that a doubling in the percentage of the population that cycle results in a 34 per cent reduction in the death rate per kilometre cycled (Jacobsen, 2003). By cycling, you will improve your own health and you will help to make cycling safer for others, encouraging more to join the growing movement.

Cycling as physical activity mainly impacts on the energy output side of the personal energy balance equation. However, because of the strong interlinkages between motorized transportation and food that we considered earlier, cycling is likely to affect your food energy intake as well. Supermarkets and petrol stations are all-purpose energy vendors. They sell cheap food energy or cheap fossil-fuel energy to whomever will buy it. Whenever you stop to fill up your car they will do their best to fill up your stomach at the same time. Cycling has the potential to uncouple personal travel from the reaches of the petro-nutritional complex.

The tentacles of the petro-nutritional complex also extend along the major public transportation routes. Wherever there are large numbers of people moving along the same transport corridors, whether by rail, bus or metro, the food environment becomes energy-dense and hazardous for personal energy balance and health. The built environment at most transportation hubs is organized such that commuters are channelled in order to provide the maximum sales opportunities for food outlets, which are often the fast-food chains. Cyclists, on the other hand, are less likely to find themselves faced with tempting fat and sugar snacks, first, because they don't need to stop off from time to time to fill the petrol tank and, second, because the spatial independence provided by cycling allows them to follow more personalized routes from their own origin to their own destination, with far fewer opportunities for commercial people channelling. And of course you can eat and smoke whilst driving a car, because this is a passive activity that does not cause you to breathe more deeply. You rarely see anyone eating or smoking whilst riding a bike. This probably at least partly explains the research finding that cycling and walking can help smokers to quit (Woodcock et al., 2009).

The main psychological obstacle to taking up cycling is that car travel has distorted our sense of distance. The main physical obstacle is that car travel has demolished our physical fitness. Cars are built for speed, and when we travel by car we imagine that we must be travelling fast and covering long distances. In fact, in most urban areas average vehicle speeds are remarkably low. Navigating a city by car is a motorized Morse code of stops and dashes. Congestion is the daily reality of car travel. Cars, whether parked or moving, take up a lot of space, which sets a limit on flow. According to research by the Worldwatch Institute, a 'one metre width equivalent right of way' can carry 170 people per hour in cars compared to 1,500 people per hour on bicycles (Worldwatch Institute, 2010). Bicycles flow in a way that cars cannot. This is reflected in average urban speeds. Data from the London Area Travel Survey shows that the average speed of cycle commuting is around 10 kilometres per hour, only slightly less than that for car commuting (11 kilometres per hour). However, we must also take into account that car users need to spend their leisure time taking exercise to stay healthy, whereas cycle users get their 'exercise' through their daily travel, journeys they make anyway.

Private passenger cars should carry a government health warning. There should be graphic images plastered on their side panels and bonnets of overweight, unfit people or clinical photographs of crud-clogged arteries. Research conducted in the sprawling, pavement-free US city of Atlanta found that each additional hour spent in a car per day is associated with a 6 per cent increase in the risk of obesity (Frank et al., 2004). Car travel, and the inactivity it entails, decimates physical fitness. The motor car is a highly effective machine for converting muscle into flab. Car use is disabling (Aldred and Woodcock, 2008). A study of trends in the aerobic fitness of children and adolescents in eleven developed countries found evidence of a rapid decline in fitness over the past twenty years as car travel has increased and active transport has decreased (Tomkinson et al., 2003). Declining physical fitness means that the speeds that people can comfortably walk and cycle are now slower, thus limiting non-motorized access. When it comes to providing access, cars give with the one hand and take with the other. In years gone by, the population would readily walk and cycle much greater distances. The English composer Sir Edward Elgar took up cycling in his forties and cycled up to 50 miles a day around the lanes of Herefordshire. He called his bicycle Mr Phoebus and wrote in his journal how cycling had inspired many of his musical works.

Ironically, it was cars that created the problem of distance in the first place. Initially, high-speed travel offered the prospect of more distant locations and shorter travel times, but soon the lower time costs of car travel made car use more attractive, leading to increased vehicle use and more congestion. Congestion has eventually resulted in reduced speeds and increased journey times. Research by David Metz at University College London has shown that the average time spent travelling has remained constant at around one hour per day for the past fifty years (Metz, 2008). Car travel creates the problem of distance, promises to solve it, but then fails.

Considering the health and environmental benefits of cycling, you might expect that promoting cycling would be a high priority for public investment. Not so. Even though obesity costs the National Health Service about £1 billion each year, in 2008 government spending on cycling in England was a paltry £10 million, about 0.1 per cent of Department for Transport spending. Meanwhile the medical profession, which you might hope would be lobbying for sustainable strategies to improve public health, remains preoccupied with drug treatments. Drugs that reduce the absorption of dietary fat, suppress appetite or increase energy expenditure make good profits for the pharmaceuticals companies. And it does not stop here. A 2008 article in the *New England Journal of Medicine* discussed the prospects of developing an exercise pill – a drug that would mimic the bodily effects of physical activity (Goodyear, 2008). The drug was proposed as a treatment for 'the great majority of Americans, probably as much as 70 per cent of the population [in whom] there is an inability or unwillingness to meet the minimum physical activity guidelines established by the American College of Sports Medicine'. Drug treatment for obesity will do nothing to reduce the threat from climate change. The NHS carbon footprint report showed that the NHS emits 21 million tonnes of carbon dioxide per year and that pharmaceuticals production accounted for about one-fifth of this total (NHS, 2009b).

The struggle over the coming years will be the struggle for road space. In high-income countries, motor vehicles have expropriated large swathes of public space. This space was taken by force, most of it in the 1920s and 1930s, a period during which there were particularly high road-death rates. For example, in 1935, there were 3,000 pedestrian deaths and 1,400 cyclist deaths in Britain, with tens of thousands of people seriously injured (DFT, 2008). During the period when the bicycle was under attack, it is likely to have been wealthy middle-aged men who inflicted the greatest

violence. At that time, the private passenger car would have been
seen as the future of elite transport. The possibility that most
families might own a car, and that many would own more than
one, must have seemed remote. The epidemic of heart disease
that would later cut down these motorized middle-aged men in
their prime was just beginning, and the concept of exercise, the
otherwise purposeless bodily movement needed to stay healthy,
had not been invented. But capitalism has its own relentless logic
and, thanks to mass production for the mass market, car travel
soon became a nightmare. Steel stallions built for speed now
stand gridlocked or nudge forward at less than walking pace. By
the end of the century cars had forced hundreds of thousands of

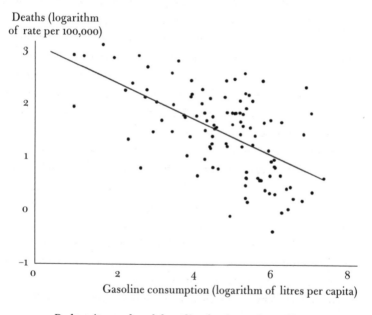

Pedestrian and pedal cyclist deaths and gasoline
consumption in 107 countries (WHO, 2009; WRI, 2010)

pedestrians and cyclists off the roads, effectively converting an epidemic of road traffic injury into a slowly unfolding epidemic of fatness and chronic disease.

On a global basis this process is ongoing. The figure on the previous page shows the association between average motor vehicle gasoline consumption and pedestrian and cyclist death rates, for the 107 countries for which the data were available (each dot represents a country). The death rates fall because, in response to road danger, pedestrians and cyclists are pushed off the streets and eventually into cars. Compare this figure with that on page 29 and you can see that as motor vehicle gasoline consumption increases, average population BMI increases with it.

The transition to cycling will represent 'development' for rich and poor countries alike. Development does not mean squandering the earth's ecological capital to maximize the financial returns of a corporate elite. Development means the expansion of human freedom within ecological bounds. Development implies freedom from hunger and disease, freedom from ignorance, freedom from violence and persecution, and freedom from the tyranny of uncontrolled fertility. For the world's poor, many of whom live dangerously close to the edge of survival, a bicycle would slash the time and energy expended getting to and from the fields and load-carrying. In this sense, a bicycle is as much a 'nutritional' intervention as food aid. By more than doubling the distance that can be covered in a given time period, bicycles could improve access to health and education. Bicycles can deliver vaccines, contraception and health-care workers; they can carry children and teachers to schools and midwives to villages.

For the wealthy, many of whom are disfigured and disheartened by the tyranny of excess body fat and enforced passivity, cycling would uncouple their daily travel from the disabling effects of petro-nutrition. Our freedom to move is one of the most simple and most magnificent. A child's stumbling first steps are invariably greeted with parental delight because they signal the acquisition of the developmental potential to enjoy this freedom. But, as any prisoner would know, the ability to move is necessary but not sufficient for freedom. There needs to be the space in which to move. Early childhood development is space exploration. Tentative forays at first but slowly increasing the radius of travel as confidence and ability grow. The expansion continues with the years, like a ripple moving out across a pond, and where ripples interact, children learn how to live in a shared social space. In the book *One False Move...*, Mayer Hillman and colleagues (2000) describe how the growth of motorization has imposed narrow confines on children's independent mobility. It describes how children have been imprisoned 'for their own safety' to protect them from road danger. Adults too have been imprisoned. The physical signs are palpable on our bodies.

Cyclists are now beginning to assert themselves. More and more cities hold car-free days when parts of the city are closed to motor vehicle traffic and the streets fill up with bicycles and pedestrians. 22 September is World Car Free Day. During such days residents get the chance to experience the absence of threat and the stillness of a street without cars. Although these days reveal the potential of the bicycle, they are not the act of rebellion that our planet needs. Personal and planetary health requires that every journey that could be cycled is cycled, every day and everywhere. Bertold Brecht wrote:

> There are those who struggle for a day and they are good.
> There are those who struggle for a year and they are better.
> There are those who struggle for many years, and they
> are better still.

But those who struggle all their lives:
These are the indispensable ones.

Faced with the conjoined twin energy disorders of climate change and pandemic fatness, our planet is entering its own special period, and just like in Castro's Cuba the bicycle will be part of its salvation. We are entering a new era of the bicycle. Cycling is enjoyable, healthy and sustainable and is the future for urban transport.

7

Reclaiming our neighbourhoods

From the second floor of the gym I can look down on the brightly coloured buses that thunder down Calle Colombia towards the centre of Medellín, and on the shoals of daffodil-yellow taxis that weave in and out between them. I can see the expressions on the faces of the passengers and they can see me. What will they make, I wonder, of a blue-eyed middle-aged man astride a stationary bike at seven o'clock on a weekday morning? For the next fifty minutes the heavily muscled instructor will clap loudly and bellow Spanish imperatives at the nine sweating cyclists. Even though he is shouting, we can only just hear him over the thumping bass of the sound system. His gaze is fixed on the second hand of the wall clock behind us, as he guides us through an imaginary landscape of uphill and down, simulated by tightening or loosening a strap around the front wheel of our motionless bikes.

If ever there was an exercise in futility it is exercise like this. Repetitive bodily movements designed to use up the excess energy that our motorized lives have destined us to accumulate. Fifty years ago 'exercise' as a concept did not exist. Physical activity meant walking or cycling, and work was remunerated activity not passivity. But motorization and a deluge of cheap oil changed everything. Once the street space had been taken over by the motoring classes, the vicious circle of increasing car use and decreasing walking and cycling became entrenched. Technically, the streets were only licensed to motorists. Number plates, which are carried only by motor vehicles and not by cyclists or pedestrians, remind us of this arrangement. Nevertheless, possession combined with brute force make up ten-tenths of the law. The battle for the street was lost.

Motorized transportation and energy-rich diets have condemned us either to the tyranny of exercise or to the indignity of obesity. Our instructor has to work hard to motivate us, because cycling on the spot, like most forms of pointless movement, is contemptibly dull. And we even have to pay for it. The expropriation of public spaces has turned those of us who can afford it into consumers of exercise in private spaces, and, since most people cannot afford to buy exercise, the accumulation of body fat can be added to the long list of socially stratified sufferings. The alternative, active travel, walking to the shops, cycling to see friends, was accessible to all before car use by the wealthy made it unpleasant and unsafe.

Doctors promote exercise as a strategy to improve health. Doctors in Britain can even 'prescribe' membership at gyms and health clubs and have even been given the responsibility of negotiating good prices for their patients. The Department of Health in Britain has issued guidance on exercise prescription schemes to make sure that patients get value for money. Obese people, who currently make up a quarter of the population in Britain, are on the list of suitable 'patients' for this 'treatment'. The prevailing view is that exercise is essential for the prevention of diseases caused by physical inactivity, including cardiovascular disease, diabetes, depression and stroke. Of course, doctors themselves are also vulnerable to the bodily effects of an environment that is stuffed with food energy but starved of the opportunities to use it, and are getting fat like the rest of society. However, unlike the general population, most doctors enjoy a hearty income. They can get to private health clubs because they own cars and can pay for them because they have money. Prescriptions for exercise fit neatly with their world-view. Whether or not doctors manage to avoid weight gain over the long term, some doctors will make money from those who fail. Obesity surgery is a rapidly growing speciality, and drugs that inhibit fat absorption are already on the market.

This book looks to a world where there is no longer any need to exercise. The only practicable response to climate change and population weight gain is that walking and cycling are re-established as the predominant modes of urban transportation. The passenger car's lease on road space has expired. It is time to reclaim the streets. Success could be measured by the number of health clubs going out of business. Change is already under way as more and more neighbourhoods are reclaimed by residents. This chapter is about how to stop exercising and start moving.

Reclaim your street

If we are going to make our streets safe again we will need the help of our neighbours. The belief that the road space outside our homes belongs to those who own motor vehicles and not to the general public is deeply rooted, and if we are going to reclaim our streets we will need political support. Strong links with our neighbours and the wider community are essential. There may be a local residents' association that we could join; if not we could start one. We need to spend more time in the street. Brushing the pavement or growing something green outside the home is one way to start. It will get us moving, it will make the street a more pleasant place to live, our presence in the street will make it safer, and we will meet people as they walk by. Picking up litter and growing plants and flowers might sound like strange recommendations, but there is more evidence behind them than for many calorie-controlled diets. A study in eight European cities found that low levels of litter and high levels of greenery were strongly associated with a lower risk of being overweight (Ellaway et al., 2005).

Spending less time on the Internet writing to people in far-away places and more time getting to know the people around us will improve our mental well-being. Shopping locally and on foot provides an opportunity to meet shopkeepers and neighbours. If you have to use a car, try parking it half a kilometre away from the home. This will ensure that you walk along your street every day, help you resist the temptation to make foolishly short car journeys, and if you walk this distance for a year you could shed a kilogram of fat.

We need to assess the level of road danger in the street. Towns are now criss-crossed by strong currents of kinetic energy and, just as the chances of drowning when crossing a river would depend on how fast the river is flowing and how deep the water is, the chances of being injured when crossing a road depend on how fast vehicles are moving (speed) and how many vehicles there are (traffic volume). Streets are dangerous if they have a large amount of kinetic (movement) energy moving along them.

We saw in Chapter 3 that motor vehicles convert chemical energy (petrol) into kinetic energy (movement). The amount of kinetic energy is given by the equation $KE = \frac{1}{2} MV^2$, where KE is kinetic energy, M is the mass or weight of the vehicle, and V is its velocity or speed. A truck travelling at 40 kilometres per hour is more dangerous than a car travelling at the same speed because the truck is so much heavier. Because velocity is squared, kinetic energy increases steeply with increasing speed; as we have seen, a doubling in vehicle speed leads to a quadrupling in kinetic energy. This explains why the risk of pedestrian death increases steeply as the speed of the impacting vehicle increases. According to studies by the UK's Department of the Environment, Transport and the Regions, when hit by a car at 40 miles per hour (64 kph), nine out of ten pedestrians die; at 30 mph (48 kph) around half die; but at 20 mph (32 kph) nine out of ten live (DETR, 1997).

Measuring traffic volume is simply a matter of counting how many vehicles pass during a given time period. Measuring speed is more difficult and requires special equipment. Your local roads department may well have speed information on your street and you should ask for it. If not, you must insist that they measure it because you need to know. You need to know the average speed because this determines the average level of risk. You also need to know the distribution of speeds in order to assess how many vehicles are travelling at dangerously high speeds. Your local authority might claim that your road is safe because no one has been injured. This is nonsense. People know road danger when they see it and they stay out of the way. They might try to go for cheap options like putting up road signs urging drivers to slow down. If you keep the focus on reducing kinetic energy, they won't be able to get away with this. They know road signs don't slow traffic.

20 mph streets save lives

The introduction of 20 mph speed zones in London streets over the last twenty years has been associated with a 42 per cent reduction in the number of casualties in these streets (Grundy et al., 2009; Bunn et al., 2003). This reduction is over and above the reduction in injuries that has occurred over time and from implementing other road safety measures. Some critics argue that introducing a 20 mph zone into a neighbourhood will only push traffic into other areas and move the road danger elsewhere. However, recent research found no evidence of 'casualty migration' to the areas adjacent to 20 mph zones, where in fact numbers of casualties fell by an average of 8 per cent.

The biggest reduction in injuries was in children younger than eleven years. This would be expected as these children are the group most likely to spend their time playing in the streets with friends immediately outside their homes. The other large effect of introducing 20 mph zones in residential streets was to halve the number of children killed or seriously injured, which might be explained by slower motor vehicle speeds. The numbers of children injured whilst cycling in the street were reduced by a quarter.

The decision about whether to turn a residential area into a 20 mph zone is usually made by the local authority. You can write to the local councillors asking for your road to be made a 20 mph zone. The council will probably consider information on the number of people (particularly pedestrians and cyclists) who have been injured, how many people walk and cycle on these roads, and the speed of the traffic. Information about the number of pedestrians and cyclists who are injured on the roads is collected by the police and is made available to the local authority. The local authority will usually send someone out to count the traffic flow and collect speed data if these are not already available.

The local authority might have prepared a plan for how it intends to reach government road safety targets. These are public documents and will provide useful contact names. They may also include useful statistical information about the numbers of people injured in your area by modes of transport and severity of injury. When analysing injury data you should be aware that many injuries are not reported to the police and that near-misses are excluded. Many roads, of course, may be so dangerous that no one dares to walk or cycle down them.

After (slowly) going through the relevant procedures, it is likely that some local authorities will decline your request to turn your street into a 20 mph zone. This is not that surprising. After a century of motorization, people who understand the need for de-motorization and who know how this can be achieved rapidly are relatively thin on the ground. There are charities and organizations dedicated to helping residents make streets safer and more livable and these will be able to offer advice and can put you in touch with like-minded individuals. But there are also influential lobby groups out there who will not like what you are doing and will try to hold you back. These are some of the arguments they will use:

- *All we need is better driver training.* In international scientific literature there have been some twenty-four randomized controlled trials of driver training, which when added together include about 300,000 drivers. A scientific review of these trials has shown that driver training is completely ineffective (Ker et al., 2005). In fact only one type of driver education has ever been shown to have statistically significant effects. Driver training in schools and colleges actually increases rates of road traffic crashes (Achara et al., 2001). There are a large number of programmes that involve going into secondary schools and talking to students who have not yet obtained a driving licence about driving and road safety. They increase road traffic crashes because students exposed to the programmes start driving sooner than they would otherwise, and the earlier they get behind the wheel the more road crashes they have. Despite clear evidence that they do more harm than good, school-based driver training remains popular and was even included in the UK road safety strategy. The car lobby loves them because they are effectively a free advertisement for driving.

- *Road humps will reduce ambulance response times.* There is reliable scientific evidence that area-wide traffic calming through the use of speed humps and cushions, chicanes and road narrowing reduces the risk of road traffic injuries by slowing down vehicles. However, one popular objection to the use of such measures is that they reduce the response times of emergency vehicles, thus putting patients at risk. Getting sick patients to hospital sooner may have a small health benefit, in which case a trade-off has to be made between the benefits of traffic calming and the drawback of slower ambulances. It is important to bear in mind here that the kinetic energy of speeding emergency vehicles is also extremely dangerous. In the UK over a five-year period, 'blue light' vehicles from the emergency services were involved in 11,925 road traffic crashes, resulting in 1,926 serious injuries and 188 deaths. Any health benefit from speeding ambulances would have to compensate for the road danger they cause. We would also need to factor in the health benefits of the increased walking and cycling that slower street speeds would encourage. Regardless, the newer designs of speed-reduction measures that do not unduly impede larger wheelbase vehicles like ambulances make the response time objection redundant.

- *We need more road safety education for children.* There is no evidence that pedestrian training for children reduces their chances of being killed or injured in traffic crashes (Duperrex et al., 2002; Akbari et al., 2001). Why, then, is pedestrian safety education so popular? The main function of road safety education is ideological. It tells children and their parents that traffic is here to stay and that unless they get out of the way they could die. The car makers worship road safety education, particularly for children in Africa and other rapidly motorizing regions. It tells the children to get out of the way.

It is vital for the motor manufacturers that road safety efforts do not get in the way of car sales in developing countries, and some road safety strategies present a serious threat. For example, reducing the need for car travel by better town planning or by building better infrastructure for walking and cycling could greatly reduce road deaths and injuries. However, such an approach would not be in the best interest of the car companies, which are facing a serious problem of manufacturing overcapacity and oversupply. Faced with the possibility that road safety could impact on prices and profits, when it came to driving forward the global road safety agenda, the car makers lost no time in taking over the wheel. Of all the departments within the WHO, the Department of Violence and Injury Prevention has the smallest budget when compared with the disease burden it is charged with tackling (Roberts et al., 2007). The car makers have been quick to exploit this weakness and have stepped in to fund various WHO road safety initiatives, thus taking effective control of the global road safety agenda. Some people would be surprised to hear that a WHO manual on speed management (GRSP, 2008) was part-funded and directed by the car makers and an offshoot of Formula One motor racing (the FIA).

Despite the best attempts by the car lobby to control the global road safety agenda, all around the world, people themselves are taking action in response to road danger. In India, almost every other day a bus or truck gets burned by local residents because it has killed a child, and if a driver does not run for his life after hitting a pedestrian then it is likely that he will get lynched. Indian villagers make it very clear that they do not accept that road deaths are unavoidable 'accidents'. Although the Indian government talks about the problem of road safety, this has not translated into an understanding of the need to reduce the level of kinetic energy on the nation's roads. Instead, it rehearses the same arguments that were made in Europe and North America during the slaughter that accompanied their rapid motorization. There are television and poster campaigns but nothing that really works. The general public, however, knows what works, and are taking action themselves with 'unofficial' road humps appearing everywhere, and not just in residential areas, but on intercity highways as well. Although these road humps are technically illegal, citizen anger is so strong that there is nothing the government can do to stop them.

Much the same is happening in Africa. In Uganda, which has the second highest road death rate in the world, people are taking to the streets to demand a reduction in road danger. The following extract is from a 2008 Uganda newspaper article:

> Tear gas would not disperse them. Nor would the firing of live ammunition by anti-riot Police. Armed with stones and other missiles, an angry mob of Nansana residents engaged anti-riot

police in running battles for over an hour, disrupting business in the busy roadside town. The residents were protesting the death of a schoolboy killed in a hit-and-run accident. He was the fourth person to get killed by a speeding vehicle within a short period at the same spot. (Kiwawulo et al., 2008)

This particular demonstration ended with residents digging up the road where the child was killed, lighting bonfires and blocking the road with tree branches and logs. 'We are tired of speeding drivers who kill innocent people', they said. Only when the local chief of police intervened and assured residents that he would personally ensure that speed humps were installed did the crowd disperse.

Streets for people

Slowing down motor vehicle traffic is the first and most important step in reclaiming the streets. Clawing back street space from motor vehicle traffic is the second. Widening pavements, installing cycle lanes and planting trees and green plants make a safer and more pleasant environment for pedestrians and cyclists and send a clear message that this is a low-energy setting. It is just as important to reclaim main roads and high streets as it is to reclaim local residential streets. Unless main roads are made safe and attractive, local shops will struggle to attract customers, and shoppers will find it more convenient to drive to the nearest supermarket. High-speed roads channel and concentrate wealth, putting many small shopkeepers out of business in the process.

Spatial planning is the key to healthy and sustainable transport and to safe, sustainable neighborhoods (Barton et al., 2009). There is no intrinsic demand for transport despite the best efforts of the car makers to convince us otherwise. What we look for is convenient access, and so where things are situated in a local community has a major impact on subsequent transportation choices. Reducing greenhouse gas emissions, promoting physically active transport, and creating safe and pleasant places to live all require that neighbourhoods are designed such that pedestrians and cyclists are given priority in accessibility planning. If walking and cycling are the easiest ways for local people to get around and to access the services that they value, then this will be reflected in their transportation choices. The focus must be on accessibility rather than mobility and this requires that communities localize as much as possible.

Sustainable spatial planning decisions require greater democracy in local and national decision-making (Barton et al., 2009). A disproportionate share of those who make decisions about the shape of our towns and cities are middle-aged men who drive to work. Unsurprisingly they tend to assume that their own transportation needs are representative of most other people's and so have designed transportation systems that give priority to cars. I was once invited to serve on the UK Department of Transport's School Travel Advisory Group. The stated objective of this group was to encourage children to walk to school, ostensibly because this would be good for their health. I was pleased and surprised that the Department of Transport should be so worried about children's health. However, it soon became obvious that their main concern was reducing congestion, which on reflection was a far more plausible explanation than were children's activity levels (Roberts, 1996). Children being driven to school were considered to be a major contributor to road traffic congestion at peak commuting times. Travel to school was a soft target that might free up road space for more important journeys. Ironically it was UK government policy that caused much of the increase in school-related car travel in the first place. Driven by the idea that parents should have greater choice when it comes to selecting schools for their children, the requirement that children attend their local primary school was relaxed. This resulted in an unholy scramble for the best schools wherever they may be located, resulting in a large increase in school-related travel distances. Because distance is an important determinant of whether a journey is made by foot or by car, this particular education policy resulted in a large increase in car use (DiGuiseppi et al., 1998). Of course, only children from car-owning families were able to enjoy the opportunities afforded by 'increased school choice'.

Despite the priority given to commuting travel, analyses of travel behaviour show that only one in five trips is related to paid employment. Shopping, leisure and personal business account for over half. Most trips are relatively short, with half not further than 5 km and 30 per cent not further than 3 km. Improving cycle access to local shops and services has huge potential to improve health and promote sustainability.

In a rapidly warming world there will be a huge demand for urban space to plant trees and plants, and road space will be prime real estate. Most roads have dark surfaces which heat up in the summer so that towns and cities become heat islands. On a clear summer afternoon the air temperature in a typical US city is between two and three degrees Celsius higher than in the surrounding countryside. As US cities have removed trees and vegetation to provide more space for roads and parking, peak urban temperatures have risen. The maximum summer temperature in downtown Los Angeles is now 2.5 degrees higher than it was in 1920 (Akbari et al., 2008). Hot cities are not only unpleasant places to live but they can also kill, and young children, the sick and the elderly are the most vulnerable. The

2003 European heatwave killed about 70,000 people, and as global warming continues the danger from such heatwaves will grow. Climate scientists estimate that the likelihood of extreme heatwaves has already doubled as a result of man-made climate change. Future similar surprises or worse are inevitable. Trees intercept sunlight before it warms buildings and roads and cool the city by evaporation. Trees and green plants can reduce maximum city temperatures by up to 3 degrees Celsius. Trees will make our streets more livable; they will reduce the danger from heatwaves, reduce the risk of flooding and urban smog, and decrease the amount of energy used in cooling homes. A study in California found that the shading provided by trees reduced the electricity used in household air conditioning by 30 per cent (Akbari et al., 2001). The above considerations and the fact that trees sequester atmospheric carbon dioxide will make them critical in adapting to and combating future climate change.

There will also be an increased demand for land to grow food. When the combustion of fossil fuels is rationed, as it will have to be in the future, and when population growth increases the demand for food, with climate change and petrol rationing decreasing the supply, the price of food will increase, particularly food that has to be transported over long distances. There is already an enormous demand for allotments, and once the street has been taken back from motor vehicles it is conceivable that some residents will want to dig up the streets.

The strategy used to reclaim the street is a creative act of the public's political imagination, as indeed will be the ways in which the street space is subsequently used. There will be many and heated arguments, but these will forge the community solidarity needed to cope with future climate chaos. You may be interested in what the residents of a neighbourhood in Leeds,

England, did when they took back control of their street. The Methleys is a neighbourhood of about 300 back-to-back tightly packed terraced houses where the only public spaces are the streets. In 1995 the Methleys Neighbourhood Action decided to take them back for the residents. Street closures for outdoor film festivals, street parties, games, sports and the locally famous 'turf the street incident' of 1996 give a sense of the possibilities. The community has its own website (www.methleys.org.uk) where the full flavour of local action can be savoured. The Methleys is now a Home Zone, a neighbourhood of streets with pedestrian priority where traffic speeds are lowered to less than 20 miles per hour and street design features make it clear that the safety of pedestrians and cyclists is what matters most. Of course, the Methleys is just a beginning.

8

Reclaim your home

This book has described the connections between fossil-fuel energy use, fatness and climate change. It has documented the central role of motor vehicle transportation in the global fatness epidemic and the close links between transportation and food in the petro-nutritional complex. Recognizing that fatness is an environmental rather than a personal problem, the middle chapters focused on the changes that will be required to make our environment both sustainable and non-obesogenic. We considered Contraction and Convergence, a public policy to ration fossil-fuel energy use, and we examined the central role of cycling and spatial planning in enabling us to live within our ration. Most of the changes necessary will require concerted political action since they relate to determinants that are beyond the control of any given individual. However, one facet of our fattening unsustainable environment over which we have considerable control is what goes on inside our homes and what and how we eat. This is the subject matter that will be covered in this chapter.

You will not find any cooking recipes in this book. Nevertheless, since this chapter is about regulating energy balance in order to control body weight, we present the main recommendations in the format of a calorie-controlled diet. After all, the calorie is a general unit of energy. It makes no difference where the energy comes from. You can work out the calories in a bag of crisps and you can work out the calories in a barrel of oil. Indeed, in their book *Food, Energy and Society*, David and Marcia Pimentel worked out that the annual food energy intake of the average American is roughly equivalent to the energy contained in a barrel of crude oil (Pimentel and Pimentel, 1996). If you eat a Big Mac with cheese, medium fries and a 12 ounce strawberry milkshake once a week for a year, then that works out in energy terms at about two and a half gallons of crude oil. It is all about energy. The difference between this chapter and a traditional diet is that this one considers more of the factors that influence energy balance than most diets. What it does have in common with most diets is that it requires commitment. But if as individuals and as a society we find it all too hard and give up, our grandchildren will never forgive us, since we will have condemned them to an uninhabitable hot, dead planet.

Our homes should be a refuge, a low-energy haven in an otherwise energy-dense world. We will have to act collectively to confront and change this wider world, but our homes we can reclaim immediately. Starting from the home, we can make sorties into a dangerous and fattening environment, like low-energy freedom fighters in search of a sustainable planet, retreating to the home from time to time to gather strength and resolve. Because we control what happens inside our homes, this step should be the easiest. However, to give us an idea about the sorts of changes that we need to make, let's consider another fattening indoor environment where we are not so free to make changes. Let's consider a night at a high-end hotel of the sort you might find in many cities around the globe.

I have sometimes been invited to speak at international medical conferences and have had the misfortune to stay in some of the best hotels in the world. Typically, having been denied the pleasurable activity of carrying my own bags and of using the stairs, I would be accelerated upwards from a glittering lobby to a carbon-copy hotel room that would light up like Times Square at Christmas as soon as the perforated plastic key was inserted into the slot by the door.

The hotel environment is designed so that guests consume as much energy as possible, whilst expending very little of their own in the process. There will be a minibar, filled with tempting chocolate bars, peanuts, crisps and other salty snacks, as well as beer, wine and spirits. There will be a wide bed facing a wide-screen television that can be controlled remotely from the bedside, along with the air conditioning, room lighting and window blinds. You can call down for food to be delivered to your room, or, if you can be bothered to get off the bed, you can go down to eat in the restaurant. Moving only your fingers you turn off the lights, close the curtains and then select from a hundred different cable television channels.

Throw out the minibar

Step one is the equivalent of throwing out the minibar. I have already explained why a fondness for fattening foods is not personal foible or a character flaw, but part of your evolutionary endowment. The desire to eat whenever food is available and the ability to store excess energy as fat is hard-wired into our biology. This is not to say that nothing has changed in the 3.5 million years since our ancestors roamed the East African savannah. For one thing, we are a little smarter then they were, thanks to our considerably larger brains. But when we come home tired after a hard day, or if some of our hard-earned extra brainpower is temporarily decommissioned after a detour to a bar, we are more like our distant relatives than we might wish. We might start searching for that cache of honey, and when we find it in our refrigerator we will eat it just like our ancestors would, and we will store the excess energy as fat, just as we are biologically programmed to do. Snacking is the modern word for foraging. Our ancestors foraged for low-energy berries and nuts but we forage for high-energy chocolate bars and biscuits. In order to maintain a healthy body-weight, in energy terms our home must resemble the African savannah. Yes there can be food in the kitchen, but low-energy food, like grains and vegetables, that take time and effort to prepare. If you cannot imagine it growing in a field somewhere, it should not be allowed into your home.

To make the necessary changes we will need the support of the people we live with. If you live with a partner, or if you have children, you will need to explain to them why changes in the home are essential. This will take time and patience but it will be worth it in the long run. If you allow energy-dense food in through the front door it will present a threat to all of you, however well it is guarded and whatever your best intentions about rationing its consumption. Nor should you offer energy-dense food as a gift to friends or work colleagues. In populations where the BMI distribution has already noticeably shifted upwards, and this means most high-income countries, doing so cannot be considered a kindness.

Fetch your own food

In the hotel environment the food appears in front of you. Because of the way the petro-nutritional complex is organized, carrying your own food is essential for personal and planetary health. The most important step is not to take the car when you go food shopping. Car use makes us fat because when we drive around we are using fossil fuels rather than food as the energy source for human movement. As we have seen, car use mostly affects energy output, unless of course we take the car to the supermarket, when it influences energy intake as well. Supermarkets know that if we shop by car, we will be able to carry more food home, and since we only buy what we can take home, they sell more food to shoppers who come by car. Supermarkets are built near major roads. They provide acres of free parking, with escalators, lifts and ramps, to ensure that we can wheel our food trolley from the checkout to the boot of the car. Most supermarket parking is free, but if not, many stores will pay for our parking provided that we spend more than a certain amount in the store. Consequently, we will buy a lot more food than we need. We will eat some of it and waste some of it. The extra food we eat will make us fat and the food that we waste will be dumped in landfill sites where the rotting vegetable matter will release methane, a potent greenhouse gas.

The physics of personal energy use means that to maintain a steady body-weight, a heavier person has to eat more food every day than a lighter person. We saw earlier that the body is a vehicle designed for personal transportation and that a heavy body is a gas (food) guzzler. Food production accounts for between 10 and 30 per cent of global greenhouse gas emissions, more than from transport or industry. The carbon footprint of a fat population is a correspondingly chubby one.

About 20 million tonnes of food are wasted in the UK each year, with household waste accounting for the largest share at about 7 million tonnes per year. British shoppers throw away about one bag of food for every three bags they buy. For most practical purposes we can take it that food is made out of fossil fuels, albeit with some help from the sun. First, energy-intensive chemical fertilizers are poured on the ground; then petroleum-powered agricultural machinery sows, waters and harvests the crop; next it is carried by trucks to factories where it is processed and packed; and then it goes on to brightly lit, centrally heated supermarkets, where the food is refrigerated and displayed. Then we come along, we buy it, we stick it in the boot of the car, we take it home and put it in the fridge. By this time our food has a carbon 'boot' print of embedded emissions, and then what do we do? We throw it away. Those greenhouse gas emissions were all in vain. Although we will not solve the problem of food waste until the food system as a whole recognizes the social and environmental costs of food production, one thing we can do to avoid buying more food than we need is to stop food shopping by car. Step two is to leave your car at home and walk to the shops.

A recipe for sustainable shopping in towns and cities goes like this. First, decide what you are going to cook over the next few days. Write down what you need on the back of an old envelope. Buying the food that you need rather than buying a range of ingredients and deciding later will reduce waste, saving you money and reducing methane emissions. Second, get your hands on a medium-sized rucksack, not huge – there is no need to look like a walking pagoda; something no higher than your shoulders and no lower than your waist will do. A rucksack avoids the wasteful use of plastic bags, protects your legs from bruises and encourages good walking posture. Third, walk around your neighbourhood shops until you find what you need. Getting to know your neighbourhood, its shopkeepers and building social networks with your local community will be essential later on when it comes to reclaiming your street. You will soon realize that it makes no sense to buy unnecessary foods because these waste space in your bag and have to be carried home. You will stop buying fizzy drinks and high-sugar fruit juices because large volumes of liquids are heavy to carry. You will buy less alcohol for the same reason. Alcohol is as energy-dense as fat. You will no longer need a huge freezer.

Food shopping will be a daily activity and to start with may take longer that driving to the supermarket. However, sooner or later sustainable spatial planning will bring the retailers closer to where you live. You will spend less time in your car, less time doing pointless exercises and perhaps less time at work, since by consuming less and wasting less you might not need to spend so much time earning the money to pay for it all. As your body gets lighter, you will spend less time eating, walking will get faster, and you will be able to move around with greater agility.

Learn how to cook

It is important to know how to cook. Unless we do, we can never escape the high-fat, high-sugar supermarket fodder that is making us fat and factory owners rich. Food produced in factories is generally higher in fat and sugar than the food that you prepare at home. This is not surprising. If you want to make a profit from selling ready-made food, it makes sound business sense to use the cheapest ingredients possible, and fat and sugar have never been cheaper. Cooking your own food will allow you to see and control how much fat and sugar go into your meals.

Ironically, one of the reasons why fat is so cheap is that your taxes have subsidized its overproduction. Until relatively recently the European Union spent about €2 billion per year in agricultural subsidies to keep milk production at levels that are about 20 per cent above domestic demand. The excess milk was converted into milk powder and butter, and since butter mountains are expensive to maintain, and politically embarrassing, the butter was sold off on the cheap to the food industry, which turned it into high-calorie cakes and puddings (Elinder, 2005). In other words, first your taxes subsidize European farmers to overproduce high-energy foods, then you solve the problem of oversupply by eating it. In terms of heart disease, butter is poison, pure and natural perhaps, but poison nevertheless. You were paying the farmers to poison you. New Zealanders eat more butter per capita than any other nationality and pay for the privilege in terms of increased rates of heart disease. Auckland University epidemiologist Professor Rod Jackson believes that butter should be taxed:

> We have a health tax on alcohol and cigarettes and there should
> be a health tax on butter. It's the most poisonous commonly
> consumed food in New Zealand. It's about the purest form of
> saturated fat you can eat and it has no protein and no calcium.
> Butter has had all the good things taken out and just left the
> poison. (Jackson, 2008)

Eat like the Japanese

The fatness pandemic has reached almost every country in the world. Some countries are affected more than others but none is out of its clutches. Our task as consumers is to consume. This is what the global economy demands of us, and if we have the money we usually obey. Of course, we must not forget that those who do not have the money to consume are not given a seat at the table. Around 850 million people remain energy-undernourished. Nevertheless, among the seated guests some are getting fatter faster than others. The USA and its economic colonies such as Mexico are the worst affected, whereas the epidemic is unfolding more slowly in Japan.

Research has been unable to establish how the Japanese have managed to keep adiposity at bay, but one difference in eating habits between the USA and Japan is portion sizes. Japanese people eat less food. They eat small portions out of small bowls. The food industry in the USA has moved heaven and earth, using all the power and influence at its disposal, to ensure that public-health messages about food and fatness focus on the type of food rather than on the amount of food eaten. 'Eat less' is the nutritional message that the food manufacturers simply cannot stomach (Nestle, 2003).

There are three basic nutritional building blocks: carbo-hydrates, proteins and fats. All foods are made up of different combinations of these basic components and all are required in a healthy diet. Anything that contains carbohydrates, proteins or fats could theoretically be part of a healthy diet. Sausages and chips contain a lot of fat but could, in moderation, form part of a healthy diet. This reasoning ensures that food makers can sell whatever they like and, having successfully suppressed any public-health messages about the amount of food you eat, their products are safe. There has been a lot of debate by health professionals about the right ratio of different food types. This ratio is often conceptualized as a food pyramid. Some foods they say should make up the wide base of the food pyramid and these should be eaten more, whereas others should be at top of the pyramid and these should be eaten less. What is not discussed is the size of the pyramid itself, and the North Americans swallow food pyramids that would put the Ancient Egyptians to shame.

So, step three is to eat like the Japanese, which means eat less. When you start walking and cycling regularly, especially for commuting, you will be able to eat a bit more, but for now you need to consume less energy by eating smaller portions. Go through your cupboards and get rid of all large plates and bowls and replace them with smaller-sized crockery. Use your old dinner plates as the serving plates for the whole family. Eat at the table, and if you want to go the whole way learn to use chopsticks. These will focus the mind and slow you down so that you start to feel satisfied before you are over-full. A study from Osaka University in Japan found that people who eat quickly or who eat until full are twice as likely to be overweight than are people who eat slowly and who stop before feeling full (Maruyama et al., 2008). People who stuff themselves and bolt down their food were three times more likely to be overweight. Mind-full rather than stomach-full is the idea. Cook enough food for the meal at hand but no more. Although in theory leftover food could be eaten the following day, it is more likely that if you cook too much you will wage war with your biological programming for the rest of the day, and I put my money on your genes winning.

Eat less animal produce

In terms of climate change, meat is an ecological disaster. Around a third of the surface of the earth is used to produce meat, either to provide pasture for grazing or to grow the grain that is used to feed the cattle. The world's forests breathe in carbon dioxide and store it away where it cannot raise the earth's temperature.

The more meat you eat, the more trees are cut down to provide the land for livestock production. Cattle are a cancer that is slowly eating away at the planet's lungs. Because grain production for cattle feed is highly fossil-fuel-intensive, meat consumption increases the amount of carbon dioxide given to the atmosphere, whilst at the same time reducing the amount that is captured by the forests. Furthermore, the methane produced from the fermentation of vegetable matter in the digestive tract of livestock is a powerful greenhouse gas.

Our ancestors enjoyed the occasional meat feast but worked for it. Unlike vegetables, meat ran away when they tried to eat it. Nowadays the catching and the killing are done for you, and on an industrial scale, and we are surrounded by dead meat. World average meat consumption is about 100 grams per person per day, the equivalent of a decent-sized beefburger every day of the week. But the global average hides huge variations depending on income. People in the USA, Canada and Europe put away two and a half burgers per day, whereas for most of Africa a burger's worth of meat every four days would be a treat (McMichael et al., 2007). Population and economic growth in poor countries are increasing the demand for meat, especially in Asia and Latin America. This will only increase the carbon footprint of cattle production.

Some people living in poor countries suffer from nutrient deficiency and could do with eating more meat. People in wealthy countries have to eat less meat to allow the others to eat more without damaging the climate. The good news is that both sides stand to benefit from this arrangement. For those who are not deficient in these nutrients, reducing the amount of animal products in the diet will reduce the risk of cardiovascular disease and bowel cancer.

Turn down the heat

The next step is to turn down the heating. Our homes are hotter than ever before. Average indoor temperatures in homes in the UK have risen from 12°C in 1970 to 18°C in 2004 and the living areas are likely to be even warmer (Utley and Shorrock, 2006). It is unfortunate that you need central heating at all. Unless you live in a particularly cold climate, a well-insulated home with double glazing should be enough to keep you warm. Turning down the heat will encourage your body to burn off some of your excess fat and will encourage you off the sofa away from the television and out into the world. Don't forget that television programmes are only made to sell adverts and most television ads encourage either gluttony or sloth (Swinburn and Shelly, 2008). Car adverts showing slim sexy people travelling at high speeds through un-spoilt countryside: does this sound like your life? The reality of car travel is overweight people, breathing in car fumes, travelling at about 18 miles per hour (the average speed of car travel in Central London), polluting the atmosphere. As regards turning down the thermostat, higher indoor temperatures may also con-tribute to food wastage. Until the 1960s most UK homes did not own a refrigerator and food was stored in a cool side room called a larder. Food expert Tara Garnet has speculated that the move towards open-plan living and greater use of central heating might have contributed to the rising popularity of the domestic fridge (Garnett, 2008). And there should be no place in your home for fossil-fuel-hungry machines that rob you of your fundamental human right to move your body. Motorized lawn mowers, hedge trimmers and that machine that blows leaves around should be banished. Such energy-saving devices consume fossil-fuel energy, contributing to climate change, and by depriving you of the opportunity to move your body – something that would build your muscle mass and increase your fitness – they leave you fatter and weaker.

Get a life

Whilst cycling across Clapham Common, I nearly rode into a man called Mark. He was busy yelling at Tarka and was not looking where he was going: 'Tarka! For goodness sake Tarka come here!' Tarka was not listening. She was sniffing another dog's bottom, wagging her tail furiously. Mark, I discovered, was a Canadian from Vancouver but was now living in London. He was joined at the waist to seven frisky dogs by a tangle of brightly coloured lanyards that were attached to his belt by a jangle of steel shackles of the sort used by climbers. As the dogs pranced around him, the red and blue lanyards splayed out in all directions and the way they came together like an umbilical cord at his navel reminded me of the blood vessels on a placenta. With his muddy boots, ruddy cheeks and lively entourage of dogs, he looked part polar explorer and part hobo.

Mark was a dog walker. He earns £15 per dog for a two-hour walk. Most of his clients are city executives. Pointing with his thumb at the parade of seven-figure properties behind him he explained 'those guys are kind of busy, they don't get the time.' Not having enough time to walk the dog presumably means they don't have enough time to walk themselves. Research in Europe, North America and Asia shows that fatness is correlated with working hours, with those working the longest hours being the fattest. Parents who work long hours are also more likely to have fat children (Foresight, 2007). Although we cannot say that long hours actually cause obesity, living in a hurry presumably makes fast food and ready-made meals more appealing, with fewer opportunities for physical activity.

Not all animals are as lucky as Mark's. Many, like their owners, are getting fat. A study in Melbourne found that 34 per cent of Australian dogs are overweight and 8 per cent are obese (German, 2006). The results for cats are very similar. According to vets at the Small Animal Hospital of the University of Liverpool,

> there is a need to increase awareness of companion animal obesity as a serious medical concern within the veterinary profession ... the problems to which obese companion animals may be predisposed include orthopedic diseases, diabetes mellitus, abnormalities in circulating blood lipid profiles, cardio-respiratory diseases, urinary disorders, reproductive disorders, neoplasia (cancer), dermatological diseases and anesthetic complications.

Fat cats in the city face a remarkably similar line-up of health risks. But it is not just top executives who are money rich, time poor and vulnerable to chronic diseases. Over 20 per cent of employed people in the UK work more than forty-five hours per week, which is high by European standards, although lower than the USA. And it is not to put the bread on the table that we work such long hours. Food has never been cheaper. According to UK government statistics we work like dogs so that we can spend our money on transport (mostly fuel for our cars), foreign holidays, sports club membership and leisure-class fees, housing and heating, food and drink (meat accounts for the lion's share) and eating out, in that order. In other words, we spend most of our money on fossil fuels (i.e. transport, food and heating) and burning off fat. Most of the stuff that we buy is packed full of embodied carbon, the emissions produced during production.

The last step in the sustainability diet is to consume less and to live a life that has purpose. There are undoubtedly myriad social and political constraints to changing the way we live and work, but if there was absolutely no element of human volition involved, there would have been no point in writing a book. However much it might serve as a metaphor for our alienation from the natural world and the things that we value, a healthy and sustainable life means far more than walking our own dog. The quest for health and sustainability demands a deeper analysis of our role in the world. As a species humankind has made some spectacular advances. We have harnessed the productive forces needed to allow the creation of leisure and art, and have developed political systems that value human rights and the full development of the individual. We have largely thrown off the leg irons of a totalitarian Church and have a scientific method that has shown the natural world to be even more remarkable than the clerics could ever have imagined. Thanks to science we understand the threat that our planet now faces. But then somehow we became slaves to the economic and social systems that had liberated us. Production–consumption became an end in itself. We consented to sell our time, our humanity and perhaps even our future in order to consume, in order to satisfy the constantly changing menu of desires that our economic system has programmed for us. If solidarity, truth and justice can be sold so readily, what hope can there be for sustainability?

9

A better world

The crisis was over. The trauma team had left the resuscitation room, leaving only the patient and a nurse, who was silently gathering up the used syringes and opened ampoules and carefully placing them in the sharps bin, taking particular care to avoid pricking herself with any blood-soiled needles. The patient was pale but his blood pressure was stable and he was conscious. A plump bag of saline to which the CRASH 2 trial treatment had been added hung from a pole by the side of his bed and was slowly being infused into his arm. A short wiry man in a dark blue pullover quietly entered the room and went over to stand by the foot of the resuscitation trolley on which the patient lay. He then pointed a gun at the patient's chest and shot him dead.

Very few bleeding trauma patients are pursued into hospital by cold-blooded assassins, as this Colombian patient was, but even in hospital patients are by no means out of danger. Despite the best medical care, the risk of in-hospital death from bleeding among the 20,000 patients included in the CRASH 2 trial was

6 per cent. It is as though every patient faces their own silent assassin, who before pulling the trigger spins the cylinder of a revolver that has the capacity for 100 deadly bullets but which contains only six.

On 10 March 2010, the blinded treatment codes were revealed to the trial staff, the analysis programme was run, and the results stuttered onto the computer screen. It had taken five years of recruitment to provide the four numbers need to calculate reliably the risk of death in the drug and placebo groups. The zeros after the decimal point in the statistical test showed that the result was real and not due to chance. The drug that we had tested significantly reduced the risk of bleeding to death. Bleeding trauma patients would still play Russian roulette with death, but one of the six bullets could now be removed from the gun. The risk of death in the drug-treated group had been reduced to 5 per cent. The trial had shown that a cheap drug that has been used for decades to treat heavy menstrual periods could save tens of thousands of lives each year (CRASH-2 trial collaborators, 2010).

The following day in the offices of the Sustainable Development Commission on London's Whitehall, health-care managers, doctors and pharmaceutical chiefs met to consider how the health sector could respond to the challenge of a low-carbon economy. Health care is a highly polluting industry. The carbon footprint of the NHS is 21 million tonnes per year, more than the total emissions of some medium-sized countries (NHS, 2010). About 60 per cent of the carbon footprint of the NHS is related to procurement – in other words, the carbon emissions from the production and distribution of the medicinal products that the NHS purchases from external suppliers. The NHS spends £20 billion on the goods and services used to treat patients, and the carbon emissions involved in their production is enormous. Most of the carbon emissions are from pharmaceuticals. Drugs are chemicals and the chemical industry is highly energy-intensive.

Is this telling us something about the choices that society will have to make? Do we have to eschew life-saving drugs, modern health care and other development benefits in order to save the planet? Is climate change mitigation the end of progress and a return to austerity? Former British prime minister Gordon Brown seemed to think so. In his October 2009 speech to the Major Economies Forum, in which he outlined the threat of unchecked climate change, he warned that 'taking this path will not be easy for any of us'. In the closing chapter of his climate book *Heat*, *Guardian* columnist George Monbiot makes a similarly grim forecast (Monbiot, 2006). After reminding us of the unequivocal science, the dark dealings of the oil industry and the connivance of governments, he pulls no punches about what a low-carbon future will entail:

> For the campaign against climate change is an odd one. Unlike almost all the public protests which have proceeded it, it is a campaign not for abundance but for austerity. It is a campaign not for more freedom but for less. Strangest of all, it is a campaign not just against other people, but also against ourselves.

This chapter presents a radically different perspective. The claim that reducing fossil fuel energy use implies self-denial and the constraint of human freedoms is mistaken. The mitigation-is-austerity slogan is nothing more than the second front in a propaganda war. The first wave of misinformation is of course climate change denial. Of the ten most powerful corporations in the world, eight are oil companies or car makers. These companies do have something to lose if we take fossil fuels out of our environment. The oil companies have responded predictably by denying that man-made climate change is real. Their tactics are crude but effective. Money and misinformation are used to sow doubt. They sprinkle cash on the public relations industry, which funds fake citizens' groups to disseminate climate-change-denial propaganda that at first sight appears independent of the oil industry. Lazy hacks then lap it up and feed the public with lies.

The second wave of propaganda is that climate change mitigation means a return to austerity. What better way to dissuade the public from taking action on climate change than to tell them how much they will suffer if they do? Come on in, the water is freezing. Eat this, it will make you sick. Burning fossil fuels has delivered important freedoms but it has also robbed us of many. It is as if we are seated on the lap of the oil industry sucking from its black nipple being told how much weaning will hurt.

According to Richard Layard from the London School of Economics, an economist who has turned the quest for population well-being into a scientific discipline, the main determinants of happiness are health, family relationships, community and friends, satisfying work, personal freedom and good values (Layard, 2006). The burning of fossil fuels and the unsustainable use of the earth's natural resources do not appear anywhere on his list of ingredients. Indeed, the New Economics Foundation has compiled a happy planet index, a measure that compares the health and well-being of a country with its ecological footprint (NEF, 2009). You may be interested to know that people living in Costa Rica live longer than people in the USA, have higher levels of life satisfaction and have an ecological footprint that is less than one-quarter the size. The unsustainable consumption of fossil fuel energy may increase corporate wealth but it is not necessary for a happy life.

A healthier, happier world

In 2009, as part of a research project to estimate the health effects of reducing fossil fuel energy use, researchers at the London School of Hygiene & Tropical Medicine estimated the health effects of transport policies that would meet greenhouse gas emissions-reduction targets (Woodcock et al., 2009). Meeting emissions targets in the transport sector will require substantial increases in walking and cycling, with correspondingly large reductions in car use. Based on the scientific evidence linking physical activity and health, it was estimated that the increase in walking and cycling would dramatically cut rates of chronic disease, with between 10 and 20 per cent less heart disease and stroke, between 12 and 18 per cent less breast cancer, and 8 per cent less dementia (Woodcock et al., 2009). Most of our carbon-laden pharmaceuticals are used to treat these chronic diseases. Because they require daily life-long treatment they generate more income for the drug industry than one-shot drugs for use in emergency situations. Indeed, the profit motive is likely to explain why the trial treatment used in the CRASH 2 trial had been tested and licensed for use in menstrual bleeding but not in traumatic or obstetric bleeding, which claims the lives of hundreds of thousands of young people each year.

The results also showed that sustainable transport would also improve our mental health, with an estimated 6 per cent less depression. It is important to note that the estimate for depression considered only the effects of increased physical activity and did not take into account the mental health benefits of more neighbourhood greenness, less community severance, reduced fatness or less noise pollution (SDC, 2010). The relation between fatness and depression works in both directions. Fatness causes depression and anxiety, and depression and anxiety lead to weight gain (Atlantis et al., 2009). Although doctors will typically diagnose 'a case of depression', in reality mood is a continuous measure ranging from robust happiness at one end to severe misery at the other. There is a population distribution of mental well-being (probably similar in shape to the population distribution of fatness considered in Chapter 2) with some very happy people in the lower tail of the distribution, some very unhappy people in the upper tail, but with most people around the average level of mental well-being. A societal increase in physical activity would shift the entire population mood distribution towards better mental health such that the overall level of social well- being would rise (Hamer et al., 2009). A 6 per cent decrease in depression really means that everyone feels a bit happier.

The London School of Hygiene & Tropical Medicine project also considered the health effects of reducing livestock production in order to limit the cattle-related methane emissions and deforestation that are contributing to global warming (Friel et al., 2009). Reducing the amount of animal products in the diet would reduce our consumption of harmful saturated animal fats, which would result in a further large fall (a 30 per cent reduction) in the incidence of chronic disease. By improving diet and physical activity levels, climate-change-mitigation policies would slash rates of diseases that mean premature death and disability for hundreds of millions of people around the world. Many of the bodily changes that accompany what we currently call ageing are merely the effects of years of exposure to dietary poisons, of which saturated animal fat is probably the most common, and a lifetime of physical inactivity. The reason why age is such a strong predictor of the risk of heart attack, stroke and dementia is that in a sedentary population fed on a high saturated-fat diet, age is a more accurate measure of our lifetime exposure to saturated fat and inactivity than is our blood cholesterol level or our body weight.

Reducing meat consumption will reduce rates of cancer of the colon and rectum. Colorectal cancer is the second most common cancer in men after lung cancer. Cutting back on animal products is good for the climate and good for health. Without the menace of heart disease and cancer and with better mental health the expectation of healthy life would increase and the amount of time spent with disability would be compressed.

Eating less animal fat and taking more physical activity would of course reduce levels of population fatness. According to the UK government, if present trends continue, by 2050, nine in ten adults will be overweight or obese. As we have seen, fatness is a disorder of fossil fuel energy abuse and there is nothing about fatness that implies freedom. Fat is a fetter that drags us into inertia and reinforces our dependence on oil.

There would be less hunger in a de-motorized world. In April 2008, Evo Morales, president of the poor and increasingly hungry Bolivia, pleaded *La vida primero los autos segundos* ('life first, cars second'), exhorting the wealthy world to stop burning food in their cars. He was objecting to Western governments' policies on biofuels for transport. However, car use and food prices were linked long before there were policies on biofuels. Car use drives up food prices because, as we have seen, oil is a key agricultural input. Gordon Brown when prime minister responded to the bio-fuels concern by calling for more agricultural research, free trade and food aid for the starving. He later went on to provide large public subsidies for the failing car industry. Reducing car use is essential to prevent starvation in poor countries. Until agriculture unshackles itself from its dependence on oil, petrol tanks in rich countries and stomachs in poor countries will be competing to be filled. The decoupling of food shopping from the petro-nutritional complex will also reduce food prices by reducing the massive food wastage that car-based food shopping promotes.

Consuming fewer animal products will also reduce food prices because cattle are fed on grain, and high meat consumption forces up world grain prices. Feeding grain to animals is a highly inefficient use of food energy in a world where millions of people go hungry. Of the £48 that the average UK household spends each week on food, £12.80 is spent on meat, £3.70 on fresh vegetables and £3.00 on fruit. Eating less meat will reduce hunger, improve health and prevent climate change. We have seen that under the framework of Contraction and Convergence there would be a transfer of wealth from the carbon-profligate world to the carbon-frugal world. This would provide people in poor countries with the economic access to food that they currently lack. People go hungry because of poverty and not a shortage of food.

The experience of Cuba in the 1990s confirms the health benefits of societal reductions in fossil fuel energy use (Franco et al., 2008). During the Cuban energy crisis the proportion of adults who were physically active more than doubled. The population average BMI fell by 1.5 units, with a halving in the prevalence of obesity from 14 per cent to 7 per cent. Deaths from diabetes fell by 51 per cent, from heart disease by 35 per cent, and from stroke by 20 per cent (Franco et al., 2007). Mass starvation was avoided, because Cubans recognize food as a human right and not an economic commodity to be rationed according to the ability to pay. Cuba ranks seventh on the New Economics Foundation Happy Planet Index for 2009. Its belligerent neighbour the USA ranks 114th, just next to Nigeria. Cuba shows that weaning off oil can be achieved whilst maintaining high levels of sustainable well-being.

A better world

Communities that value humanity

Low-energy towns and cities could be better places to live. There would be less motor vehicle traffic, and what traffic there would be would entail lighter vehicles travelling at much slower speeds. Traffic today blights our environment in the way that open drains blighted Victorian towns and cities. Two hundred years ago our streets were awash with sewage and our cities stank. Infectious diseases left thousands of first birthdays uncelebrated, but it was the stench not the health problems that finally drove politicians to put our sewage underground. Joseph Bazalgette was commissioned to build an underground system for the disposal of London's waste. The sewers did more to improve health and well-being than any other public policy that century. Getting kinetic energy out of our streets could be the next great advance. We can look to a future where there will be fewer road deaths and injuries, cleaner air and much less noise. The urban infrastructure would show a new respect for humanity. The web of social networks between citizens that has been swept aside by raw kinetic energy will slowly be rewoven.

Walking and cycling and public transport will be the predominant modes of urban travel. China has shown that cycling can provide an effective means of mass transportation. Until the early 1990s, the streets of China were thick with black and silver Flying Pigeon bicycles, wide rivers of low-energy human movement, unhurried and safe (Roberts, 1995). Stripped of its wide border of cars, the canvas of public space will be fully rolled out for the public to paint on. There will be more people out in the streets, more children playing, more community activity, more art and more green space. Green space has been shown to improve both physical and mental health. The benefits for physical health are likely to be due to increased walking in greener areas, whereas the restorative effects of natural environments account for the mental-health benefits. The real promise of low-carbon transportation is that cities and towns will be more attractive and enjoyable places to live and work. The city will be a quieter place, and, without its coat of combusted carbon, a brighter, more colourful place.

We saw in Chapter 2 that the main correlates of increasing BMI during the 1970s and 1980s were increasing car use and increased television viewing. The average Briton now watches three and a half hours of television per day. Car use and television viewing are related. Road danger has contributed importantly to the wholesale retreat into the home, particularly in the case of children, although the pull factor of passive entertainment is likely to have played a part. According to Layard, television viewing decreases levels of happiness by raising our standards of comparison. We are besieged by images of people who are wealthier, healthier and slimmer than we are. And the more time we spend passively watching, the more weight we gain and the greater the gap between our own body shape and the shapes of the people we watch. The commercial objective of television is to encourage consumption. Every hour spent watching television results in an extra US$4 spent on consumption (Layard, 2006). If the de-motorization of society means more activity, better health, less fat, stronger community and less promoted consumption, then we have little to lose and much to gain from reducing fossil fuel energy use.

Recessions, depressions and car use

In August 2007, the global financial system went into crisis. The major banks had spun a cobweb of financial complexity to conceal their fraudulent lending practices, and when the webs were swept aside to reveal the extent of the liabilities the global financial system went into meltdown. The credit bubble that the bankers had inflated over the previous two decades had burst in their faces. Years of easy money had resulted in an unserviceable financial debt and an unsustainable ecological debt. In the boom years, conspicuous overconsumption of fossil fuel energy had been all the rage. In 2007, energy-thirsty sports utility vehicles accounted for 29 per cent of new vehicle sales in the USA, with SUV sales in the UK rising rapidly. In the USA, two out of every three motor vehicles sold were bought on credit. Cheap money had set the scene for the worst economic recession since the 1970s. All that was needed was an oil price rise to tip the global economy into recession. When it finally arrived in 2008, the economic consequences were predictably severe.

Almost every major economic recession in the USA has been preceded by an oil price rise, and every oil price rise has been followed by a recession. Although the USA produces about 5 million barrels of oil per day, it consumes 19 million barrels per day, 13 million of which it burns on its highways. Due to its car dependency the US economy is hugely vulnerable to oil price rises. Ironically, it was the low petrol prices in the 1990s and early 2000s that inflated the credit bubble in the first place. Low petrol prices kept inflation down, which in turn made it possible to keep interest rates low. However, between January 2004 and January 2006, US oil prices surged from $35 to $68 per barrel. This set off a burst of inflation which pushed interest rates up from 1 to 5 per cent. Suddenly paying the mortgage and the interest on the loans for the two family 4x4s became a lot more difficult, quite apart from the costs of filling them up at the pumps. Food prices also went up. Then the bubble burst, the banks went bust and the economy went into recession.

Britain, Europe and Japan were just as vulnerable. Transport accounts for the largest share (13.4 per cent) of household spending in the UK, with nearly half going to fuelling and servicing the family vehicle. When petrol prices rise, people have less money to spend, and, since oil is an ingredient in nearly all consumer goods, what is on sale in the shops is also more expensive. Recreation is the second leading UK expenditure category, one-quarter of which is on package holidays, most of which are outside the UK. Unsurprisingly, Spain has been hit hard by the recession.

When the financial firestorm began, the immediate response from government was to get unsustainable consumption back on the roads. In both Europe and the USA, car makers were given cheap loans that cost the taxpayer billions. European car makers claimed that the loans would be used to manufacture greener vehicles. In 1998, the car makers had promised to cut the greenhouse gas emissions of their vehicles voluntarily. By 2006, it was clear they had lied and the European Commission announced that it would set compulsory standards. By 2009, they were claiming that they could only go green with massive public subsidies.

In the UK, the government introduced a 'vehicle scrappage scheme'. This involved public subsidies to replace old cars for newer more fuel-efficient ones. Although most new vehicles produce lower carbon dioxide emissions per kilometre, the emissions involved in producing and scrapping vehicles also has be taken into account. Reducing the average lifetime of a motor vehicle increases car production, which means more emissions from car assembly. Research by the Dutch National Institute of Public Health and the Environment found that when the lifecycle emissions of vehicle production and destruction are taken into account, scrappage schemes will probably increase greenhouse gas emissions (Van

Wee et al., 2000) . This increase will be even greater if owning a newer vehicle means that people travel further or faster, something that is not at all unlikely. The vehicle scrappage scheme is another public subsidy for the petro-nutritional complex.

Following the Great Depression in the USA in the 1930s, government intervention was critical in limiting the social damage from the mass unemployment that followed. President Roosevelt's New Deal comprised huge public works that helped to boost employment and stimulate business. Armies of workers were enlisted to build the highways and bridges that later paved the way for the spectacular rise of the American automotive industry. Investment in infrastructure would also create new jobs today, but pandemic fatness, climate change and an increasing demand for a decreasing supply of oil mean that any new infrastructure would have to lay the foundations for a low-carbon society. Propping up the car industry through public loans and by subsidizing the purchase of new cars is not the solution we need.

What is needed is an ambitious decarbonization programme that will cut across all the major areas of fossil fuel energy use. This would include the decarbonization of our energy supplies, increasing the energy efficiency of our homes, the creation of an urban infrastructure for safe walking and cycling and the greening of our cities. Renovating towns and cities for walking and cycling will require architects, artists, arboculturists, builders, carpenters, engineers, ecologists, educators, planners, planters, street performers and urban farmers. Their job description would be to ensure that walking and cycling provide the most enjoyable, the most satisfying, the safest and the most direct means of getting around. These will be socially useful jobs that build the foundations for better health, safer and stronger communities and a sustainable economy. Resuscitating a carbon-based economy is a short-term fix that can only fail. As soon as the economy starts growing and the demand for oil increases, petrol price rises will choke it back into a recession. Decarbonization is not the path towards austerity – it is the only way to avoid it.

A world with less poverty

Each year worldwide, some 10 million children die before their fifth birthday. Whereas Victorian England did not know how to prevent the industrial slaughter of children from infectious diseases, today we do, yet 10 million children still die. They die because they are poor. Poverty condemns them to hunger, dirty water, poor sanitation, inadequate health care and illiterate parents. The wealthy world's response to this outrage is charity. International aid organizations have to put on increasingly elaborate media extravaganzas in order to press world leaders into giving, but their commitment is fickle.

A large part of the wealth that rich nations now hand out as charity was accumulated by expropriating the human and natural resources of the poor nations during centuries of colonial exploitation, a practice that continues today under the banner of transnational trade. The continuing haemorrhage of resources from South to North was, and still is, made possible by the exploitation of fossil fuel energy. As a result, the wealthy world has left the planet with a large environmental debt.

Contraction and Convergence is a policy response to climate change that simultaneously ensures ecological sustainability and justice for the poor. The carbon-profligate will have to pay the carbon-thrifty for their unused rights to the atmosphere, thus assuring that the limits of environmental tolerance are not exceeded. Because in general the rich lead high-carbon lives, and the poor lead low-carbon lives, the policy of Contraction and Convergence could lead to a massive global redistribution of wealth. The lives of African children should not depend on the charitable whims of the wealthy world. Rich nations can and should return some of the wealth they have taken from Africa over centuries of colonial exploitation. Funding for health care, HIV prevention, contraception and education is urgently needed. Indeed, one major opportunity for improving health whilst mitigating climate change that has not been considered in this book is greater access to family planning. This is not about controlling anyone's fertility but about providing all women in the world with the same access to contraception that rich women currently enjoy. Access to contraceptives reduces maternal death by reducing the total number of births but especially births at the extremes of the reproductive period, very young mothers and older mothers. Increasing the spacing of births also reduces infant mortality and malnutrition. Funding health and education in Africa should be called reparation not aid. Contraction and Convergence ensures that African children's rights to the atmosphere are bought from them rather than stolen. Tony Blair and the Commission for Africa were wrong when they stated that more roads will allow Africa's economy to prosper. Loans for road building will increase car use by the wealthiest sections of society but will do little for the majority. Contraction and Convergence changes the direction of policy from aid for road building to payments to Africans for their unused carbon rations.

A world with less war

Few events in the last thirty years have sparked more open public outrage than the war in Iraq. Nevertheless, in many respects the war was inevitable. It could be said that the blueprint for the war was set down by American town planners in the mid-1920s. That it would be in Iraq was decided by George Bush and his oil industry cronies. The architects of the war were not military planners but town planners. The United States had paved itself into a corner. Its infrastructure had become so highly car-dependent that America had become pathologically addicted to oil. More than two-thirds of US oil consumption is for transportation. Because it takes 13 million barrels of oil per day to keep the US transportation system moving, the US military has no option other than to roam the world plundering oil.

In 1997, a Carnegie Commission report on the prevention of deadly global conflict identified the factors that send nation-states to war. Securing access to essential resources was at the top of the list. The USA needs oil like a junkie needs heroin. Sprawling suburban America had developed an insatiable thirst for cheap oil, and a brutal dictator called Saddam Hussein was in charge of a country with about 112 billion barrels, which was then considered to be the largest supply in the world outside Saudi Arabia. Over 100,000 civilian deaths later, Iraq is in bloody chaos and the threat of extended global conflict has grown. Weaning transportation systems off fossil fuels could be less painful than feeding the habit.

Heroin addiction leads to crime and violence and oil addiction leads to corruption and war. In 2003, a Christian Aid report, *Fuelling Poverty: Oil, War and Corruption*, highlighted how oil 'wealth' is a curse for the countries that have it. Nigeria, Africa's most populous state, had 36 billion barrels of proven oil reserves as of 2007 and is Africa's largest oil producer. Nevertheless, one-third of Nigeria's children are underweight and more than half are without access to clean water. Oil money allows politicians to remain in power by doling out backhanders rather than building a stable economy. Easy access to oil revenues takes away the incentive to develop transparent governmental structures. With oil money gushing from the ground there is no need to regulate and tax a productive economy and so the nation-state remains underdeveloped. The list of oil-cursed countries whose governance is blighted by corruption includes Angola, Azerbaijan, Chad, Ecuador, Indonesia, Iran, Iraq, Kazakhstan, Libya, Nigeria, Russia, Sudan, Venezuela and Yemen. The volatile mix of poverty and corruption inevitably leads to violence. Reducing the global demand for oil through de-motorization would mean that the oil stays safely in the ground.

And then there is Afghanistan. Opinion polls show that both the American and the British public are confused about the reasons for the war in Afghanistan. They cannot be blamed. Foreign bombardment is not the most obvious way to reduce the threat of terrorist attack, and it is unusual for Western governments to sacrifice troops to increase female education. The US Army is the military wing of the US Department of Energy, and wherever there is a US military presence there is likely to be a strong geostrategic link with energy. Afghanistan might not have oil reserves but neighbouring Iran does – 139 billion barrels. There are also extensive energy reserves around the Caspian Sea area which cannot easily be brought to global markets other than through Afghanistan. With increasing global motorization, the demand for petrol will soon outstrip supply and prices will rise steeply. American motorists will feel these price increases, which will hurt them, economically and physically. Not only is America's infrastructure oil-dependent, its population have become oil-dependent too. The average BMI of the US population is such that for many people even walking short distances is an unpleasant physical experience. When petrol prices rocket, they will demand government action, even if it is a short-term solution and even if it means going to war. At the moment the stage is being set for war on Iran. Of course, unless the world commits to rapid decarbonization, accelerating climate change may also bring new reasons for war. An increase in the frequency and severity of droughts, and decreased runoff from glaciers, will threaten the water supplies of many heavily populated areas, leading to increasing violence between states and regions. Petrol is not the only essential resource. A lack of food and water can also send nation-states to war.

A mass movement

The world trade system runs on cheap transport. Cheap transport allows business to take advantage of the low wages of poor people in poor countries and to take advantage of their natural resources. Because materials have to be transported to factories and because manufactured goods have to be transported to markets, cheap transport has reduced the price of everything. There was never a time when there was so much affordable stuff in the world. We have lived through a period of unrivalled consumerism.

However, we have yet to pay for our century of consumption. We have incurred a massive and potentially un-payable ecological debt. We have set in train a series of climatic positive feedbacks that threaten the viability of life on earth. Underpinning the climate crisis is an array of lesser vicious circles, all of which are fuelled by low-priced transport. Cheap transport has brought us the increased population fatness and decreased population fitness that have increased our dependence on motorized transport. It has increased road danger for pedestrians and cyclists, driving them off the streets and into cars, again increasing the demand for transport. It has designed into our daily lives distances that can only be conquered with low-priced motor vehicle transport. It has made controlling oil supplies the primary strategic objective of nation-states so that scarce resources that should be devoted to building a sustainable economy are instead spent on war and destruction.

The decarbonization of transport could lead us to a better world, but it will not be achieved without struggle. Success depends on the demise of some of the most powerful corporations in the world, which will resist any change using every weapon at their disposal. They will disseminate lies, blackmail governments, bribe opinion leaders and drag sustainability efforts through the courts. But there is no longer any alternative. We have to reclaim the streets and reassert our right to move. The de-motorization of our streets will require the greatest human mobilization in history.

Despite their simplicity and their scientific base, the arguments made in this book are unlikely to be given air to breathe in our day-to-day political discourse. This book has claimed that fatness is a political problem, that petrol and not food is making us fat, that 'traffic accidents' are state-sanctioned violence, that we should stop exercising, that mouths and petrol tanks are competing to be filled, and that much of what is called private profit is really public subsidy. These ideas would be considered heretical by most mainstream politicians. Nor could such claims be made in the media. Newspapers would never report on the monstrous anger of the trucks and on the pedestrians and cyclists who die like cattle, because to do so their correspondents would first have to perceive it as such, and they do not because it has been obscured by what Noam Chomsky has described 'the veil of distortion and misrepresentation, ideology and class interest, through which the events of current history are presented to us'. In the struggle for a healthy and sustainable planet, this thick fog of distortion is every bit as dangerous as the physical threats it serves to conceal. For without the confusion provided by a relentless drizzle of lies, the car and oil companies could never have created such a colourful rainbow of delusion from a brutal machine and barrel of oil.

References

Achara, S., Adeyemi, B., Dosekun, E., Kelleher, S., Lansley, M., Male, I., Reynolds, L., Roberts, I., and Smailbegovic, M. (2001) 'Evidence based road safety: the Driving Standards Agency's schools programme', *Lancet* 358: 230–32.

Akbari, H., Pomerantz, M., and Taha, H. (2001) 'Cool surfaces and shade trees to reduce energy use and improve air quality in urban areas', *Solar Energy* 70: 295–310.

Aldred, R., and Woodcock, J. (2008) 'Transport: challenging disabling environments', *Local Environment* 13: 485–96.

allAfrica (2009) 'Road crash casualties hit maternal health efforts', 12 June, http://allafrica.com/stories/200906120769.html.

Atlantis, E., Goldney, R., and Wittert, G. (2009) 'Obesity and depression or anxiety', *BMJ* 339: 3868.

Baffes, J. (2007) 'Oil spills and other commodities', *Resources Policy* 32: 126–34.

Banister, D., Wright, L., and Transport Research Laboratory (2005) *The Role of Transport in Supporting Sub-national Growth: An Overview Study*, London: DFID.

Barnett, A. (2005) 'Big bonuses go to rulers of aid empire', *Observer*, 22 September.

Barton, H., Grant, M., And Insall, P. (2009) 'How to help plan a healthy sustainable low carbon community', in J. Griffiths (ed.), *The Health Practitioner's Guide to Climate Change,* London: Earthscan.

Bassett, D., Pucher, J., Buehler, R., Thompson, D., and Crouter, S. (2008) 'Walking, cycling and obesity rates in Europe, North America, and Australia', *Journal of Physical Activity and Health* 5: 795–814.

BBC (2000) 'Models link to teenage anorexia', 30 May, http://news.bbc.co.uk/1/hi/769290.stm.

Bell, A.C., Ge, K., and Popkin, B.M. (2002) 'The road to obesity or the path to prevention: motorized transportation and obesity in China', *Obes Res* 10(4): 277–83.

Blair, T. (2008) 'Tony Blair on breaking the climate deadlock', http://tony-blairoffice.org/2008/06/foreword-breaking-the-climate.html.

Blundell, J., and Gillett, A. (2001) 'Control of food intake in the obese', *Obes Res* 9(4): 263S–270S.

Bunn, F., Collier, T., Frost, C., Ker, K., Roberts, I., and Wentz, R. (2003) 'Traffic calming for the prevention of road traffic injuries: systematic review and meta-analysis', *Inj Prev* 9: 200–204.

CRASH-2 trial collaborators (2010) 'Effects of tranexamic acid on death, vascular occlusive events, and blood transfusion in trauma patients with significant haemorrhage (CRASH-2): a randomised, placebo-controlled trial', *Lancet* 376: 23–32.

DEFRA (2008) *Synthesis Report on the Findings from Defra's pre-feasibility Study into Personal Carbon Trading,* London: Department for Environment, Food and Rural Affairs.

DETR (1997) *Road Safety Strategy: Current Problems and Future Solutions,* London: Department of the Environment, Transport and the Regions.

DFID (2009) 'Clearing a path for transport links in Africa to boost trade', DFID press release, London.

DFT (2007) 'Shopping', personal travel factsheet, July, London: Department for Transport.

DFT (2008) *Road Casualties Great Britain* 2007, Annual Report, London: Department for Transport.

DiGuiseppi, C., Roberts, I., Li, L., and Allen, D. (1998) 'Determinants of car travel on daily journeys to school: cross sectional survey of primary school children', *BMJ* 316: 1426–8.

Duperrex, O., Bunn, F., and Roberts, I. (2002) 'Safety education of pedestrians for injury prevention: a systematic review of randomised controlled trials', *BMJ* 324: 1129.

References

Edwards, P., and Roberts, I. (2008) 'Transport policy is food policy', *Lancet* 371: 1661.

Edwards, P., and Roberts, I. (2009) 'Population adiposity and climate change', *International Journal of Epidemiology* 38: 1137–40.

Edwards, P., Roberts, I., Green, J., and Lutchmun, S. (2006) 'Deaths from injury in children and employment status in family: analysis of trends in class specific death rates', *BMJ* 333: 119.

Egger, G. (2007) 'Personal carbon trading: a potential "stealth intervention" for obesity reduction?', *Med J Aust* 187: 185–7.

Egger, G. (2008) 'Dousing our inflammatory environment(s): is personal carbon trading an option for reducing obesity – and climate change?', *Obes Rev* 9(5): 456–63.

Egger, G. (2009) 'Should obesity be the main game? Or do we need an environmental makeover to combat the inflammatory and chronic disease epidemics?', *Obes Rev* 10(2): 237–49.

Elinder, L.S. (2005) 'Obesity, hunger, and agriculture: the damaging role of subsidies', *BMJ* 331: 1333–6.

Ellaway, A., Macintyre, S., and X., B. (2005) 'Graffiti, greenery, and obesity in adults: secondary analysis of European cross sectional survey', *BMJ* 331: 611–12.

EU (2006) 'Environmental impact of products (EIPRO): analysis of the life cycle environmental impacts related to the total final consumption of the EU25', European Commission Technical Report EUR 22284 EN.

Foresight (2007) *Tackling Obesities: Future Choices – Modelling Future Trends in Obesity and Their Impact on Health*, 2nd edn, London: Government Office for Science.

Franco, M., Ordunez, P., Caballero, B., and Cooper, R. S. (2008) 'Obesity reduction and its possible consequences: what can we learn from Cuba's Special Period?', *CMAJ* 178: 1032–4.

Franco, M., Ordunez, P., Caballero, B., Tapia Granados, J.A., Lazo, M., Bernal, J.L., Guallar, E., and Cooper, R.S. (2007) 'Impact of energy intake, physical activity, and population-wide weight loss on cardiovascular disease and diabetes mortality in Cuba, 1980–2005', *Am J Epidemiol* 166: 1374–80.

Frank, L., Andresen, M., and Schmid, T. (2004) 'Obesity relationships with community design, physical activity, and time spent in cars', *American Journal of Preventive Medicine* 27: 87–96.

Friel, S., Dangour, A. D., Garnett, T., Lock, K., Chalabi, Z., Roberts, I.,

Butler, A., Butler, C. D., Waage, J., Mcmichael, A. J., and Haines, A. (2009) 'Public health benefits of strategies to reduce greenhouse-gas emissions: food and agriculture', *Lancet* 374: 2016–25.

Fugate, D. (2007) 'Neuromarketing: a layman's look at neuroscience and its potential application to marketing practice', *Journal of Consumer Marketing* 24: 385–94.

Galeano, E. (1973) *Open Veins of Latin America*, New York: Monthly Review Press.

Garnett, T. (2008) *Cooking Up a Storm*, Centre for Environmental Strategy, University of Surrey.

GCI (1990) *Contraction and Convergence*, Global Commons Institute, www.gci.org.uk/briefings.html.

German, A. (2006) 'The growing problem of obesity in dogs and cats', *Journal of Nutrition* 136: 1940S–1946S.

Goodyear, L. (2008) 'The exercise pill – too good to be true?, *New England Journal of Medicine* 359: 1842–4.

GRSP (2008) *Speed Management: A Road Safety Manual for Decision-makers and Practitioners*, Geneva: Global Road Safety Partnership.

Grundy, C., Steinbach, R., Edwards, P., Green, J., Armstrong, B., and Wilkinson, P. (2009) 'Effect of 20 mph traffic speed zones on road injuries in London, 1986–2006: controlled interrupted time series analysis', *BMJ* 339.

GTZ (2009) *Fuel Prices and Taxation with Comparative Tables for 160 Countries*, Escborn: Deutsche Gesellschaft für Technische Zusammenarbeit.

Haile, F. (1989) 'Women fuelwood carriers and the supply of household energy in Addis Ababa', *Canadian Journal of African Studies* 23: 442–51.

Hamer, M., Stamatakis, E., and Steptoe, A. (2009) 'Dose response relationship between physical activity and mental health: the Scottish Health Survey', *Br J Sports Med* 43: 1111–14.

Harrington, S. (2008) 'The role of sugar-sweetened beverage consumption in adolescent obesity: a review of the literature', *Journal of School Nursing* 24: 3–12.

He, M., and Evans, A. (2007) 'Are parents aware that their children are overweight or obese?', *Can Fam Physician* 53: 1493–9.

Hillman, M. (2004) *How We Can Save the Planet*, London: Penguin.

Hillman, M., Adams, J., and Whitelegg, J. (2000) *One False Move...*

References

Study of Children's Independent Mobility, London: Policy Studies Institute.

Hillman, M., and Fawcett, T. (2005) *A Brief Introduction to Personal Carbon Rationing,* Edinburgh: UK Energy Research Centre.

IFAD (2008) *High Food Prices: Impact and Recommendations,* Rome: International Fund for Agricultural Development.

Jackson, R. (2008) 'Professor calls for tax on "poison" butter', 6 April, www.stuff.co.nz/national/349682.

Jacobs, J. (1992) *The Death and Life of Great American Cities,* New York: Random House.

Jacobsen, P. L. (2003) 'Safety in numbers: more walkers and bicyclists, safer walking and bicycling', *Inj Prev* 9: 205–9.

Jerrett, M., Mcconnell, R., Chang, C., Wolch, J., Reynolds, K., Lurmann, F., Gilliland, F., and Berhane, K. (2010) 'Automobile traffic around the home and attained body mass index: a longitudinal cohort study of children aged 10–18 years', *Prev Med* 50: S50–8.

Johnson, F., Cooke, L., Croker, H., and Wardle, J. (2008) 'Changing perceptions of weight in Great Britain: comparison of two population surveys', *BMJ* 337: a494.

Ker, K., Roberts, I., Collier, T., Beyer, F., Bunn, F., and Frost, C. (2005) 'Post-licence driver education for the prevention of road traffic crashes: a systematic review of randomised controlled trials', *Accid Anal Prev* 37: 305–13.

Kiwakulu C. (2010) 'Nansana residents riot over accidents', *Sunday Vision,* 21 July.

Lauderdale, D.S., and Rathouz, P.J. (2000) 'Body mass index in a US national sample of Asian Americans: effects of nativity, years since immigration and socioeconomic status', *Int J Obes Relat Metab Disord* 24: 1188–94.

Layard, R. (2006) *Happiness: Lessons from a New Science,* London: Penguin.

Lobstein, T., and Jackson-Leach, R. (2007) 'International comparisons of obesity trends, determinants and responses', www.foresight.gov.uk.

Ludwig, D., and Nestle, M. (2008) 'Can the food industry play a constructive role in the obesity epidemic', *JAMA* 300: 1808–10.

Manson, J.E., Willett, W.C., Stampfer, M.J., Colditz, G.A., Hunter, D.J., Hankinson, S.E., Hennekens, C.H., and Speizer, F.E. (1995) 'Body weight and mortality among women', *N Engl J Med* 333: 677–85.

The energy glut

Maruyama, K., Sato, S., Ohira, T., Maeda, K., Noda, H., Kubota, Y., Nishimura, S., Kitamura, A., Kiyama, M., Okada, T., Imano, H., Nakamura, M., Ishikawa, Y., Kurokawa, M., Sasaki, S., and Iso, H. (2008) 'The joint impact on being overweight of self reported behaviours of eating quickly and eating until full: cross sectional survey', *BMJ* 337: a2002.

McGreal, C. (2009) 'UK's $1bn transport network across Africa', *Guardian*, 20 February: 26.

McLaren, L. (2007) 'Socioeconomic status and obesity', *Epidemiol Rev* 29: 29–48.

McMichael, A., Powles, J., Butler, C., and R., U. (2007) 'Food, livestock production, energy, climate change, and health', *Lancet* 370: 1253–63.

Metz, D. (2008) '*The Limits to Travel: How Far Will You Go?*, London: Earthscan.

Monbiot, G. (2006) *Heat: How to Stop the Planet Burning*, London: Penguin.

NEF (2009) *Happy Planet Index*, London: New Economics Foundation, www.happyplanetindex.org.

Nestle, M. (2003) *Food Politics: How the Food Industry Influences Nutrition and Health*, Berkeley: University of California Press.

NHS (2009a) *Change4Life*, www.nhs.uk/change4life.

NHS (2009b) *NHS England Carbon Emissions: Carbon Footprinting Report* 2008, Cambridge: NHS Sustainable Development Unit.

NHS (2010) *Saving Carbon Improving Health: Update NHS Carbon Reduction Strategy*, NHS Sustainable Development Unit.

NYCPCC (2009) 'Climate risk information', February, New York: New York City Panel on Climate Change.

Olds, J., and Milner, P. (1954) 'Positive reinforcement produced by electrical stimulation of septal area and other regions of rat brain', *Journal of Comparative Physiology and Psychology* 47: 419–27.

Peden, M., Scurfield, R., Sleet, D., Mohan, D., Hyder, A., Jarawan, E., and Mathers, C. (2004) *World Report on Road Traffic Injury Prevention*, Geneva: World Health Organization.

Peters, D. (2000) 'Gender and transport in developing countries', background paper in preparation for CSD-9 London/Berlin UNED Forum/German Federal Environment Ministry.

Peters, D. (2001) *Breadwinners, Homemakers and Beasts of Burden: A*

172

Gender Perspective on Transport and Mobility, London: Sustainable Development International.

Pimentel, D., and Pimentel, M. (1996) *Food, Energy and Society*, University of Colorado, Department of Fine Arts.

Prentice, A., and Jebb, S. (1995) 'Obesity in Britain: gluttony or sloth?', *BMJ* 311: 437–9.

Riverson, J.D.N., and Carapetis, S. (1991) *Intermediate Means of Transport in Sub-Saharan Africa: Its Potential for Improving Rural Travel and Transport*, Washington DC: World Bank.

Roberts, I. (1995) 'Letter from Chengdu: China takes to the roads', *BMJ* 310: 1311–13.

Roberts, I. (1996) 'Safely to school', *Lancet* 347: 1642.

Roberts, I. (2003) 'Car wars', *Guardian*, 18 January.

Roberts, I. (2004) 'Injury and globalisation', *Injury Prevention* 10: 65–6.

Roberts, I. (2005) 'Death on the road to international development', *BMJ* 330: 972.

Roberts, I. (2007) 'Formula One and global road safety', *Journal of the Royal Society of Medicine* 100: 1–3.

Roberts, I. (2008) 'Corporate capture and Coca-Cola', *Lancet* 372: 1934–5.

Roberts, I., Carlin, J., Bennett, E., Bergstrom, B., Guyer, B., Nolan, T., Norton, R., Pless, I., Rao, R., and Stevenson, M. (1997) 'An international study of the exposure of children to traffic', *Injury Prevention* 3: 89–93.

Roberts, I., and Coggan, C. (1994) 'Blaming children for child pedestrian injuries', *Soc Sci Med,* 38: 749–53.

Roberts, I., Hosord, T., and Edwards, P. (2001) 'The World Health Organization and the prevention of injuries: phone book analysis', *BMJ* 323: 1485.

Roberts, I., Marshall, R., and Norton, R. (1992) 'Child pedestrian mortality and traffic volume in New Zealand', *BMJ* 305: 283.

Roberts, I., Norton, R., Jackson, R., Dunn, R., and Hassall, I. (1995) 'Effect of environmental factors on the risk of injury of child pedestrians by motor-vehicles: a case control study', *BMJ* 310: 91–4.

Roberts, I., Wentz, R., and Edwards, P. (2006) 'Car manufacturers and global road safety: a word frequency analysis of road safety documents', *Inj Prev* 12: 320–2.

Rodney, W. (1981) *How Europe Underdeveloped Africa*, Washington DC: Howard University Press.

Rose, G. (2001) 'Sick individuals and sick populations', *Int J Epidemiol* 30(3): 427–32.

SDC (2010) *Sustainable Development: The Key to Tackling Health Inequalities*, London: Sustainable Development Commission.

Seeney M. (2009) 'Government in £275m anti-obesity drive', *Guardian*, 2 January.

Sieber, N. (1997) 'Appropriate transport and rural development: economic effects of an integrated rural transport project in Tanzania', *World Transport Policy and Practice* 3.

Smith, R. (2002) 'In search of "non-disease"', *BMJ* 324: 883–5.

Stamatakis, E., Zaninotto, P., Falaschetti, E., Mindell, J., and Head, J. (2010) 'Time trends in childhood and adolescent obesity in England from 1995 to 2007 and projections of prevalence to 2015', *J Epidemiol Community Health* 64(2): 167–74.

Stealth of Nations (2009) 'Hawkers wanted', 16 February, http://stealthofnations.blogspot.com/2009/02/hawkers-wanted.html.

Stiglitz, J., and Bilmes, L. (2008) *The Three Trillion Dollar War*, New York: W.W. Norton.

Story, M., and French, S. (2004) 'Food advertising and marketing directed at children and adolescents in the US', *International Journal of Behavioral Nutrition and Physical Activity* 1(3).

Swinburn, B., and Egger, G. (2004) 'The runaway weight gain train: too many accelerators, not enough brakes', *BMJ* 329: 736–9.

Swinburn, B., And Shelly, A. (2008) 'Effects of TV time and other sedentary pursuits', *Int J Obes (Lond)* 32(7): S132–6.

Tomkinson, G.R., Leger, L.A., Olds, T.S., and Cazorla, G. (2003) 'Secular trends in the performance of children and adolescents (1980–2000): an analysis of 55 studies of the 20m shuttle run test in 11 countries', *Sports Med* 33: 285–300.

Truong, K.D., and Sturm, R. (2005) 'Weight gain trends across sociodemographic groups in the United States', *Am J Public Health* 95: 1602–6.

Utley, J., and Shorrock, L. (2006) 'Domestic energy factfile: owner-occupied, local authority, private rented and registered social landlord homes', DEFRA Building Research Establishement, Energy Saving Trust.

References

Van Wee, B., Moll, H., and Dirks, J. (2000) 'Environmental impact of scrapping old cars', *Transportation Research* D: 137–43.

Wang, D., Roberts, I., Tu, Y., Wang, Y., and Wu, X. (1995) 'Increasing road deaths in rural China', *J Traffic Med* 23: 129–30.

Whitlock, G., Lewington, S., Sherliker, P., Clarke, R., Emberson, J., Halsey, J., Qizilbash, N., Collins, R., and Peto, R. (2009) 'Body-mass index and cause-specific mortality in 900 000 adults: collaborative analyses of 57 prospective studies', *Lancet* 373: 1083–96.

WHO (2004) *World Report on Road Traffic Injury Prevention*, Geneva: World Health Organization.

WHO (2009) *Global Status Report on Road Safety: Time for Action*, Geneva: World Health Organization.

WHO (2010a) Global Infobase, World Health Organization, http://apps.who.int/infobase/report.aspx?iso=AFG&rid=111&goButton=Go.

WHO (2010b) *Projections of Mortality and Burden of Disease, 2004–2030*, World Health Organization, www.who.int/healthinfo/global_burden_disease/projections/en/.

Woodcock, J., Banister, D., Edwards, P., Prentice, A., and Roberts, I. (2007) 'Energy and transport', *Lancet* 370: 1078–88.

Woodcock, J., Edwards, P., Tonne, C., Armstrong, B., Ashiru, O., Banister, D., Beevers, S., Chalabi, Z., Chowdhury, Z., Cohen, A., Franco, O., Haines, A., Hickman, R., Lindsay, G., Mittal, I., Mohan, D., Tiwari, G., Woodward, A., and Roberts, I. (2009) 'Public health benefits of strategies to reduce greenhouse-gas emissions: urban land transport', *Lancet* 374: 1930–43.

Worldwatch Institute (2010) 'Matters of Scale – Bicycle Frame', www.worldwatch.org/node/4057.

WRI (2010) *EarthTrends*, http://earthtrends.wri.org/searchable_db/index.php?theme=6&variable_ID=292&action=select_countries.

Wu, Y. (2006) 'Overweight and obesity in China', *BMJ* 333: 362–3.

Index

Index